Mathematics and Physics of Neutron Radiography

Reidel Texts in the Mathematical Sciences

A Graduate-Level Book Series

Mathematics and Physics of Neutron Radiography

by

A. A. Harms
Department of Engineering Physics, McMaster University,
Hamilton, Ontario, Canada

and

D. R. Wyman
Cancer Research Foundation,
Hamilton, Ontario, Canada

Springer-Science+Business Media, B.V.

Library of Congress Cataloging in Publication Data

CIP

Harms, A. A.
Mathematics and physics of neutron radiography.

Bibliography: p.
Includes index.
1. Neutron radiography. I. Wyman, D. R. (Douglas R.), 1957–
II. Title.
QC793.5.N4628H37 1986 621.36'7 85-31239
ISBN 978-94-015-6939-2 ISBN 978-94-015-6937-8 (eBook)
DOI 10.1007/978-94-015-6937-8

Originally published by D. Reidel Publishing Company, Dordrecht, Holland in 1986
Softcover reprint of the hardcover 1st edition 1986

TO: U.M.C., P. & W.W.
E.A.B., C.R. & F.J.W.

Table of Contents

Preface

Neutron radiography has in recent years emerged as a useful and complementary technology for radiation diagnosis. It is now routinely used in industrial quality assurance and in support of selected research and developmental activities. Conferences are held on the subject, pertinent handbooks exist, and technical papers appear regularly reporting on new developments.

While neutron radiography has indeed passed through the transition from a scientific curiosity to technological relevance, it is a sign of its continuing dynamic evolution that little material has appeared which provides an integrated mathematical and physical analysis of the subject possessing both an instructional as well as reference function. It is our hope that this monograph will fill this need.

The distinctiveness of neutron radiography rests on the unique interactions between neutrons and nuclei. This leads to some special relationships between the material and geometrical properties of an object and the neutron radiographic image. The evolution of a technical discipline demands that specific conceptual constructs be developed and their mathematical representations examined and compared with controlled experiments. Experience has convinced us that a particular and substantial body of knowledge has accumulated endowing neutron radiography with the essential foundations of a unique mathematical and physical science.

Our scientific and professional involvement in neutron radiography began some 15 years ago when the senior author (A.A.H.) found himself with convenient access to the McMaster University Nuclear Reactor and research support from the Government of Canada. Over the years a number of specific research projects had been undertaken generally with the aim of elucidating particular processes and establishing appropriate mathematical formulations. Graduate students were drawn in, some academic colleagues developed complementary interests, several commercial radiographic contracts were supervised, and contact with the international community of neutron radiography undertaken. This period of time served well to integrate various perspectives, synthesize new constructs, and in general seek an intrinsically coherent characterization of neutron radiography. The coauthor (D.R.W.) joined the venture in recent years first as a graduate student and then as associate. What is contained here then, is the cumulative effect of years of involvement in a scientific-technological undertaking which demanded a significant part of our research time.

In planning this monograph, we have been conscious of the large body of related literature on neutron physics and medical-industrial radiography. Indeed, neutron radiography may be viewed as a direct descendant of these two technologies and -- as offspring are apt to do -- has introduced unique characteristics of its own. The unique characteristics are the requirements for neutron conversion in order to produce a radiograph and the role of neutron scattering in the final image. These distinctions of neutron radiography form the main themes we have sought to develop in this monograph in a manner which is both complementary to the existing literature on neutron radiography while providing an integrative perspective with a pedagogical orientation. Of necessity then, a mathematical emphasis supported by graphical aids has been chosen as the main tool of exposition. Indeed, considerable emphasis has been focussed on providing a clear linkage between physical processes and their mathematical description. Additionally, we have sought to direct our analysis specifically to provide clarification and/or direction in a form and context to aid both the beginning researcher in and the experienced practitioner of neutron radiography.

Our neutron radiographic endeavors have been much facilitated and enhanced by a number of graduate students who took a research interest in the subject; in the order of their appearance in our research group this included G.R. Norman, P.S.W. Chan, T.G. Blake, S.F. Ng, L.D. Molson, M.S. Moniz, G.S. Okawara, D.M. Lowe, W.J. Hartmann, M.P. Butler and J.C. Osuwa. Several of our academic colleagues interacted with us on specialized aspects of the subject: Professors J.H. Crocket, B.K. Garside, T.J. Kennett, and N.E. Wilson. Encouragement for our experimental work has been provided by the McMaster Nuclear Reactor supervisory staff of Peter Ernst and Jack McDougall and more recently, Don Smith. It is, however, with a particular sense of satisfaction that we acknowledge our stimulating contacts with an international community of scientists who shared similar interests; among these we cite in particular Rene Barbalet (France), John Barton (U.S.A.), Harry Berger (U.S.A.), Jozef Domanus (Denmark), Poul Gade-Nielsen (Denmark), Peter von der Hardt (The Netherlands), Mike Hawkesworth (England), Manfred Heindler (Austria), André Laporte

(France), Howard Larson (U.S.A.), H.P. Leeflang and Joachim Markgraf (The Netherlands), Ron Matfield (Italy), J. Michel and J.P. Perves (France), Jozeph Rant (Yugoslavia), Helmut Rauch (Austria), Heike Reijonen (Finland), Wade Richards (U.S.A.), Alan Ross (Canada), Gad Shani (Israel), Bill Whittemore (U.S.A.) and Anton Zeilinger (Austria). All have influenced our thinking about and working in neutron radiography and we express our appreciation to them.

The figures in this monograph have been prepared at the McMaster University Audio-Visual Department, typing was undertaken at the Engineering Word Processing Centre, and some of the calculations were done by S. D'Alessio. David Larner from Reidel Publishing provided expert editorial counsel.

To all we express our thanks.

November 1985
Hamilton, Canada

A.A. Harms
D.R. Wyman*

*Present affiliation: Ontario Cancer Foundation, Hamilton, Canada.

I. Introduction

Neutron radiography possesses important features that distinguish it from other imaging techniques. We consider these in order to place the subject into its historical and scientific-technological context.

1.1 HISTORICAL EVOLUTION

Radiography has its roots in the pioneering discovery of Roentgen (B.1)* who, in 1895, obtained a radiographic image using a high-voltage vacuum tube and fluorescent screen. The implications of this discovery of X-rays were soon employed in medical diagnosis and, in due course, a new domain of medical practice emerged. Scientists subsequently made use of X-rays as a probe for the study of fundamental properties of matter and technologists used this radiation for inspection of manufactured products. The eventual availability of γ-ray sources provided a further expansion of diagnostic uses of electromagnetic radiation.

It was, however, Chadwick's seminal discovery of the neutron (B.2) in 1932 which led to important new directions in science and technology. Unlike X-rays and γ-rays, neutron interaction is characterized by nuclear rather than electronic characteristics of the medium through which it passes. The property that neutrons constitute a non-ionizing form of radiation -- together with the recognition that neutron production was more difficult than X-ray production -- posed a number of practical problems for their applications in a radiographic context.

The earliest experimental demonstration that neutrons were of radiographic relevance was provided in a series of experiments by Kallman in the 1930's (A.2, A.11)*. However, the more broadly based development of neutron radiographic applications had to await the advent of sufficiently intense neutron beams which became available with the development of research reactors in the 1950's.

---

* These identifiers refer to Bibliography and References.

Beginning in the late 1950's, neutron radiography experienced a steady expansion in scientific development and technological applications. The reactor installations at Harwell, England, and at the Argonne National Laboratory, U.S.A., served a primary function in this process. Subsequently, research reactors at universities and research centres contributed to the dissemination of neutron radiography knowledge and its applications. Several publications (A.8,A.43,A.46,A.51,A.55,A.58) have been selected specifically to provide an indication of existing neutron radiographic installations, current practice and applications, as well as ongoing research.

1.2 ESSENTIAL FEATURES

Neutron radiography involves three principal components:

a) a suitable neutron beam,

b) an object of radiographic interest,

c) a device to record -- promptly or latently -- the radiation intensity information associated with the neutron beam transmitted through the object.

Figure 1.1 provides a graphical depiction of these three system components for the case of a non-radioactive object specimen.

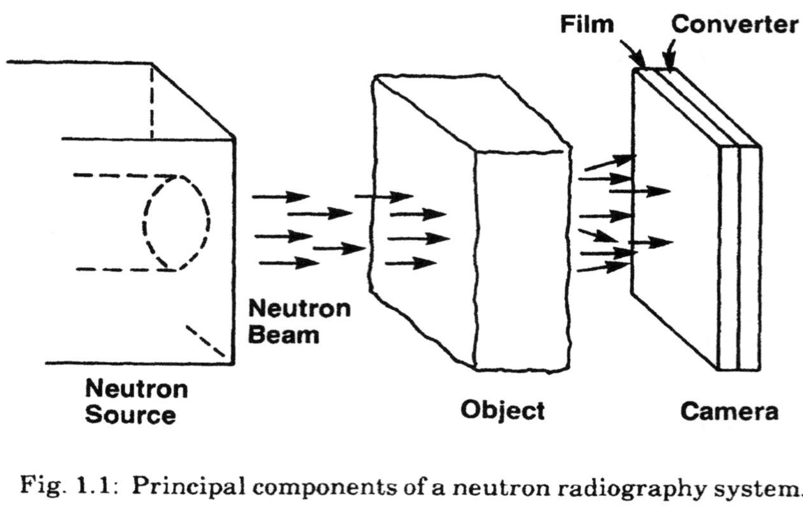

Fig. 1.1: Principal components of a neutron radiography system.

Neutron beams are most commonly supplied by nuclear fission reactors. Alternative neutron sources can be provided by accelerators and radioisotopes. The important considerations involve the intensity of the neutron source, the spectrum of neutron energies, and the collimation of the neutron beam.

As neutrons pass through an object, they interact with nuclei by scattering and absorption. The probability of such events is an isotopic, rather than elemental, property and accounts for the unique radiographic information available with neutron beams. We depict this feature in Fig. 1.2, which shows the attenuation parameters for various materials exposed to a thermal neutron beam; the important distinction between neutron absorption and neutron scattering will be further discussed in subsequent chapters.

Since neutrons constitute a non-ionizing form of radiation, they do not directly lead to the formation of developable centers in conventional radiographic film. This necessitates the placement of a converter chosen for its efficiency to capture neutrons and emit appropriate radiation. A range of choices are possible for this purpose. Metallic foils which emit internal conversion electrons, alpha particles, γ-rays, or light-emitting scintillation screens are frequent choices. The associated recording film must match the radiation emitted. Commercially available X-ray film or cellulose track-etch recorders are commonly used. High quality imaging requires that close contact between the converter and the film recorder be established, thus requiring the use of vacuum cassettes.

References A.55 and A.58 should be consulted concerning recent experience with various imaging devices.

Figure 1.1 illustrates what is frequently called "direct" neutron radiography and is characterized by an object which is not radioactive. Should, however, the object be sufficiently radioactive, a conventional film recorder cannot be used during the neutron exposure because the radiation from the object would blacken the film, thereby masking information encoded in the transmitted neutron beam. Neutron radiography of such radioactive objects then involves a converter characterized by delayed emission of radiation upon neutron capture. That is, in this case of "indirect" neutron radiography the transmitted neutron beam

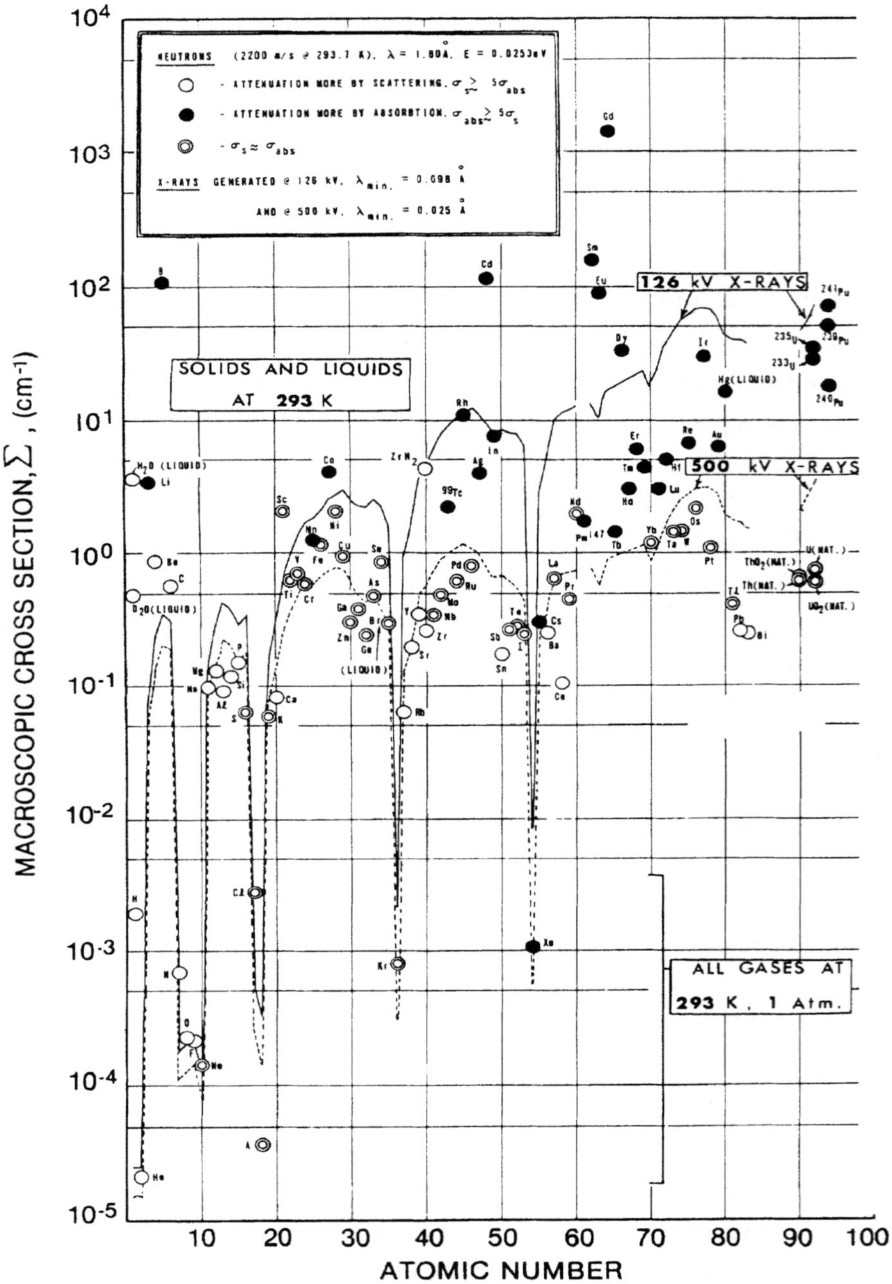

Fig. 1.2: Macroscopic neutron-nucleus cross section for various materials; X-ray linear attenuation parameter is shown for comparison (source: A. Ross, Atomic Energy of Canada Ltd.).

renders the converter radioactive and this image is extracted by subsequently placing the converter in contact with a film-recorder, Fig. 1.3. This last imaging stage is appropriately referred to as "autoradiography".

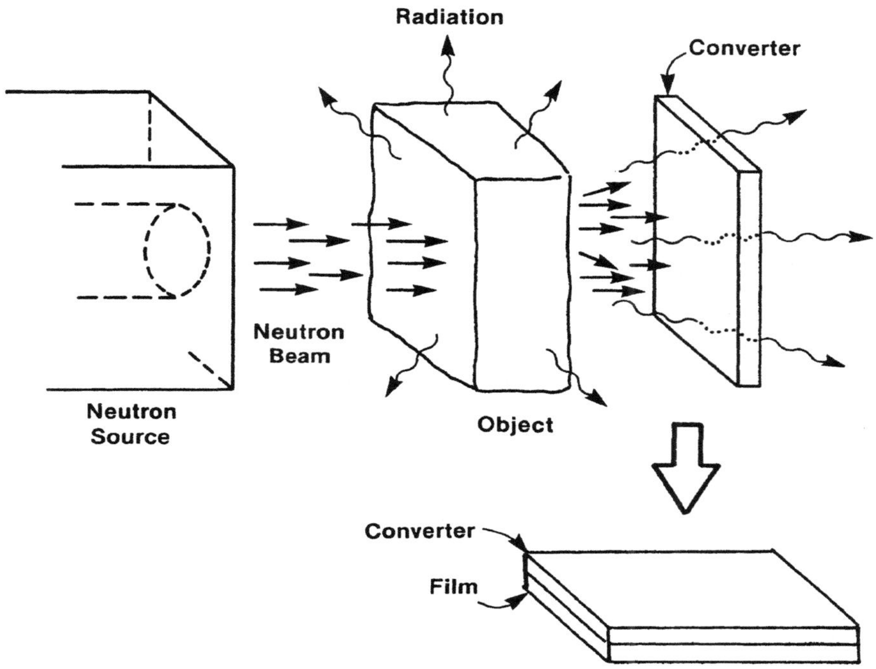

Fig. 1.3: Graphical depiction of indirect neutron radiography.

Both theoretical considerations and experimental demonstration have confirmed that neutron radiography is particularly well suited for the following two classes of radiographic applications:

A. Examination of radioactive objects which would normally blacken radiographic film. This involves the method of indirect neutron radiography and thus provides a radiographic image which cannot be obtained by X-ray or γ-ray radiography.

B. Examination of objects with particularly important features suggested by their neutron attenuation parameters:

a) detection of low density materials,

b) penetration through certain high density materials,

c) distinguishing isotopic variations.

These considerations have shown that neutron radiography does provide for unique and/or enhanced methods of radiographic diagnosis.

1.3 CONTEXT AND SCOPE

In order to provide a mathematical-physical basis of the various processes depicted in Fig. 1.1, we introduce the conceptual and graphical construction of Fig. 1.4. Here, the z-axis

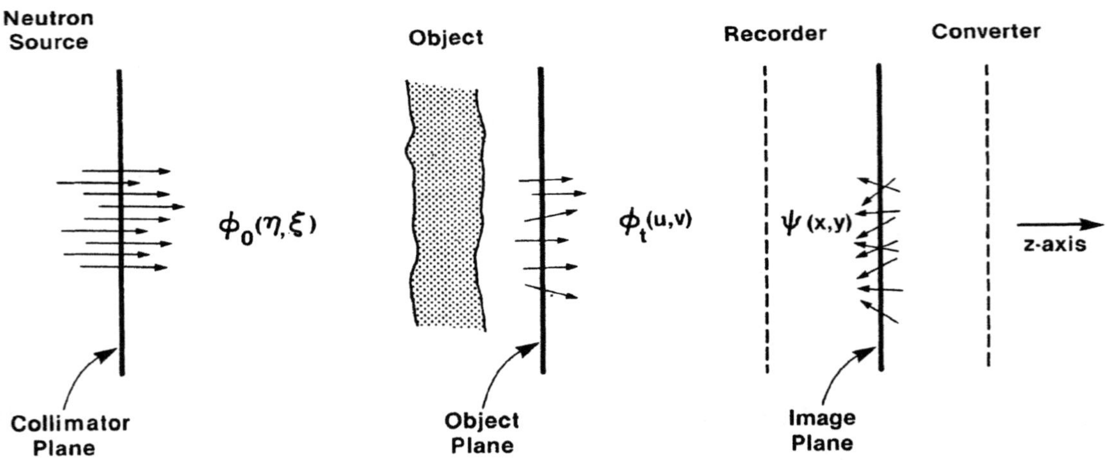

Fig. 1.4: Illustration showing the association of various radiation flux quantities with functional planes and system components.

points in the direction of the neutron beam emerging from the neutron source and is intercepted by three parallel planes characterized by orthogonal axes (η,ξ), (u,v), and (x,y) which serve the following functions:

(η,ξ): The "collimator" plane is associated with the neutron beam from a collimator device; the function $\phi_0(\eta,\xi)$ on this plane will generally represent the spatial distribution of neutrons which are directed towards the object to be radiographed.

(u,v): The "object" plane is associated with the object of radiographic interest; the associated function $\phi_t(u,v)$ will generally represent the spatial distribution of the neutron beam after its passage through the object.

(x,y): The "image" plane is associated with the neutron radiographic image; the function $\psi(x,y)$ describes the film blackening secondary radiation emerging from the converter and incident upon the recorder. For the case of indirect neutron radiography, this function will be the secondary radiation integrated over the time the converter is in contact with the recorder.

The functions $\phi_0(\eta,\xi)$ and $\phi_t(u,v)$ each represent a flux of neutrons crossing some planar surface area in a unit time interval and are expressed in units of cm^{-2}-s^{-1}. The former is fully determined by the characteristics of the neutron source and its collimator device while the latter includes the attenuation effects of the object. Hence, $\phi_t(u,v)$ is a function of $\phi_0(\eta,\xi)$ and the object.

The function $\psi(x,y)$ represents a flux of a certain type of radiation induced by the neutrons from the $\phi_t(u,v)$ beam captured in the converter. Hence $\psi(x,y)$ is also expressed in units of cm^{-2}-s^{-1} and is a function of $\phi_t(u,v)$. This then makes $\psi(x,y)$ a function of both $\phi_0(\eta,\xi)$ and the object.

However, the end result of a neutron radiographic exposure is usually the optical density in a film, the track density of a cellulose recorder, or the amplified distribution of an electronic current signal. This resultant image on the (x,y) plane, denoted by the optical density function $D(x,y)$ if photographic X-ray film is used, varies in a determinable way with $\Psi(x,y)$.

In summary, then, our emphasis will be on the elucidation of the sequential process

$$\phi_0(\eta,\xi) \rightarrow \phi_t(u,v) \rightarrow \psi(x,y) \tag{1.1}$$

for a set of objects of neutron radiographic interest. This constitutes a basic analysis because it seeks to address the fundamental and distinguishing question of neutron radiography: how do the several processes combine to transform a given physical object into a radiographic image? A knowledge of the relevant phenomena and the application of suitable mathematical constructs then promises the extraction of useful information about the object from the image.

Much of the following analysis concerns the processes that lead to some degradation of the neutron radiographic image. Indeed, considerable effort is currently directed to the avoidance or correction of image degradation, much of which arises because image formation is a "neighborhood" process. That is, the image of a point of interest invariably contains additional information from other points within some neighborhood of that point. Extending from this consideration of point image degradation are several important mathematical tools of imaging analysis. These tools can be used not only to describe image degradation but also to effect image restoration. The general problem of image restoration is outside the scope of this text. However, local restoration involving the extraction of precise information from neutron radiographs is here discussed and analyzed because it follows directly from the description of image degradation. It should be noted that conventional image restoration techniques, such as linear filtering and algebraic restoration (B.4), can provide substantial visual image enhancement, but may yield unpredictable results if precise dimensional information is to be obtained from the enhanced image.

The digital approach to image processing (B.4,B.11) has well developed applications in image restoration. Presently, however, most neutron radiography installations proceed only with analog imaging and, accordingly, we here present neutron imaging analysis using continuous, rather than discrete, methods of mathematical analysis. Likewise, most neutron radiography is currently conducted using thermal neutrons, although a few applications exist for imaging using neutrons with energies outside the thermal range. While the material here presented is intended for thermal neutron radiography, its validity extends to applications in which energy independent interaction parameters well characterize the neutron scattering and absorption processes.

Finally, the subject of neutron tomography involves the complete three-dimensional reconstruction of an object from multiple angle neutron images. The science of reconstruction from projections is well developed and documented (B.3). We will thus restrict our attention to the processes involved in obtaining, and the analysis of, single neutron radiographs.

II. Component Characterization

The beam, object, converter and recorder constitute the basic components of the neutron radiographic system. Relevant physical processes for each component may be identified, described and analyzed.

2.1 COLLIMATED BEAM

The ideal collimated neutron beam is suggested in Fig. 2.1. It is frequently possible to take for the collimator function, $\phi_0(\eta,\xi)$ of Fig. 1.4,

$$\phi_0(\eta,\xi) \simeq \phi_c \qquad (2.1)$$

where ϕ_c is a constant. This parameter is frequently called the neutron beam intensity, the neutron scalar flux or simply the neutron flux. Physically, it represents the number of neutrons per unit area per unit time passing through a suitably defined area; its common units are neutrons/cm^2–s. The direction of motion of all neutrons is normal to area A_c and parallel to the positive direction of the z-axis, Fig. 2.1. Typical values of ϕ_c for thermal neutron radiography using a reactor neutron source are $\phi_c > 10^6$ n/cm^2–s.

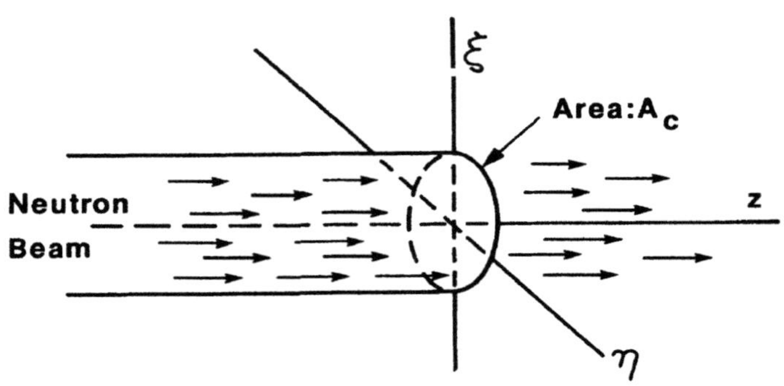

Fig. 2.1: Depiction of a collimated neutron beam.

2.2 DIVERGENT BEAM

We next consider an isotropic source of thermal neutrons incident on a surface area A_S. Using methods of optical ray tracing, a divergent neutron beam can be characterized as shown in Fig. 2.2. The validity of this representation and ensuing analysis rests on the assumption that negligibly few source neutrons scatter from the collimator wall through the collimator area A_c and that, similarly, no significant number of neutrons enter the beam through the collimator wall.

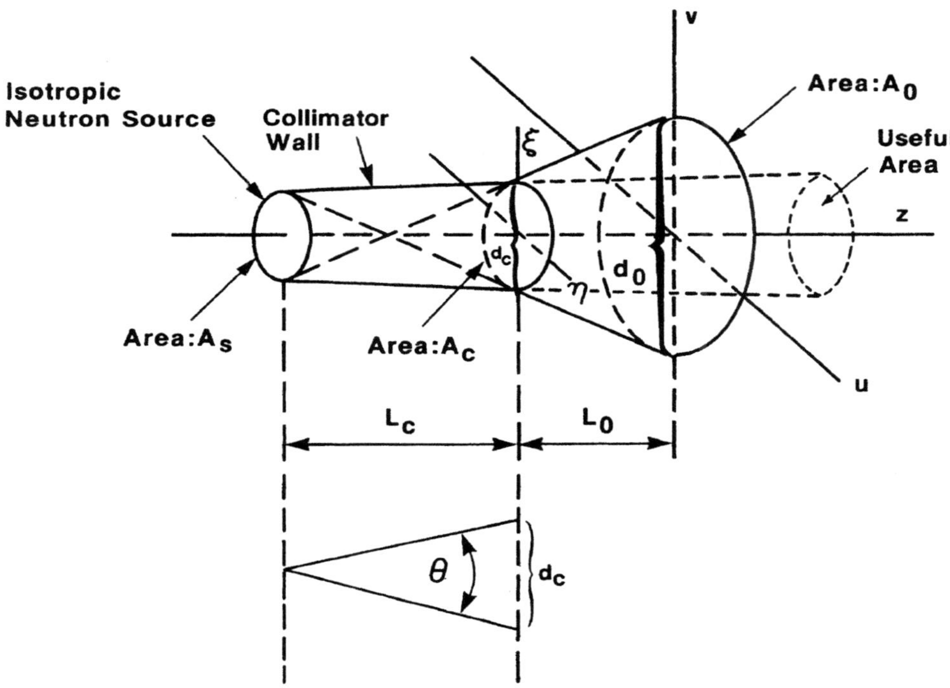

Fig. 2.2: Geometric parameters associated with a divergent neutron beam.

The neutron flux exiting at the collimator area A_c can be taken to be a known constant so that, as for Eq. (2.1),

$$\phi_o(\eta,\xi) \simeq \phi_c \tag{2.2}$$

Beam divergence may be defined in several ways, for example, by a ratio such as

$$R_\theta = \frac{L_c}{d_c} \tag{2.3}$$

which relates to the angle of divergence by

$$\theta \simeq \frac{1}{R_\theta} \tag{2.4}$$

becoming exact in the limit of large L_c. Beam divergences of $R_\theta > 100$ are frequently encountered with reactor neutron sources.

With the collimator beam ϕ_c as a measurable quantity and all geometric parameters of Fig. 2.2 known, some important collimator plane (η,ξ) and object plane (u,v) relationships can be established. For example, starting with neutron conservation, we write the approximate expression

$$\phi_c A_c \simeq \phi_o A_o \tag{2.5}$$

where the subscripts refer to the collimator plane and object plane. The scalar neutron flux at the object plane is thus

$$\phi_o \simeq \phi_c\left(\frac{A_c}{A_o}\right) \tag{2.6}$$

which, for circular or square apertures, reduces to

$$\phi_o \simeq \phi_c\left(\frac{d_c}{d_o}\right)^2 \tag{2.7}$$

with d_o and d_c the respective dimensions of the object and collimator planes.

The above divergent beam description leads to an inverse square relationship between the neutron scalar flux at the collimator and object planes. This relationship rests on the approximate neutron conservation expression in Eq. (2.5), in particular, on the assumption of spatially constant flux, ϕ_c and ϕ_o. On the other hand, the collimated beam description of Sec. 2.1 leads to equal collimator and object plane flux. In reality, the neutron beam intensity reduction upon exiting the collimator plane lies between these two extremes.

For many practical cases of interest, the area A_c extended to the object plane (u,v) is important because the neutron beam is likely to possess maximum uniformity over this area at the object plane, as indicated in Fig. 2.2.

2.3 ABSORBING OBJECT

Consider a collimated beam of neutrons impinging on an absorbing object as suggested in Fig. 2.3. An ideal neutron absorbing object is taken to be one for which neutron absorption is the dominant process; that is, scattering interactions are taken to be sufficiently rare relative to absorption that they need not be considered. Hence, the neutrons in a beam will either be absorbed by a nucleus or continue to pass through the object without collision.

Fundamental beam interaction considerations suggest that the fractional decrease in a neutron beam per unit distance in the beam direction is a constant for the absorption process. That is, the number of neutrons absorbed is proportional to the number of neutrons present and to the extent of the absorbing domain. We write this plausibility statement at an arbitrary z-coordinate in the medium as a differential expression in terms of the neutron flux as follows

$$d\phi/\phi(z) = -\Sigma_a(z)\,dz \tag{2.8}$$

Here, the macroscopic absorption cross section at a z-coordinate, $\Sigma_a(z)$, is the proportionality constant. This parameter is related to the various isotopes of the medium by

$$\Sigma_a(z) = \sum_i \sigma_{a,i} N_i(z) \tag{2.9}$$

where $N_i(z)$ is the number density of the ith type isotope at a z - coordinate of interest and $\sigma_{a,i}$ is its associated microscopic cross section (B.10) appropriate to the neutron energy spectrum of the beam. The minus sign in Eq. (2.8) implies a decrease in the neutron flux, $d\phi$, per unit positive distance increment, dz, with $\Sigma_a(z)$ as a positive quantity.

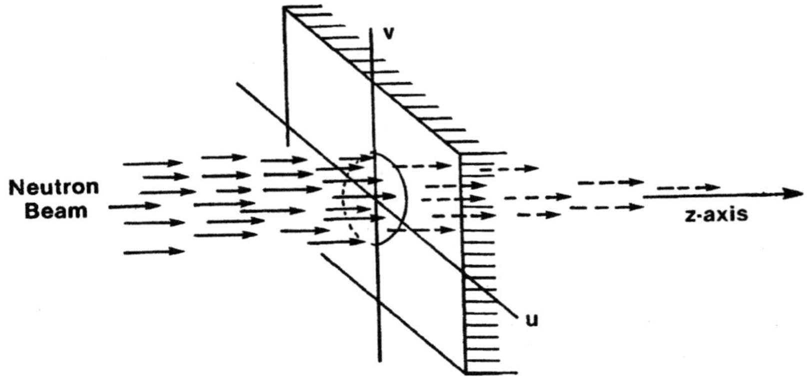

Fig. 2.3: Attenuation of a collimated neutron beam in an absorbing object.

Equation (2.8) constitutes a first-order differential equation

$$\frac{d\phi}{dz} = -\Sigma_a(z)\,\phi(z)\,, \qquad \phi(0) = \phi_c \tag{2.10}$$

where we also specify the boundary condition of the flux with the z-coordinate axis chosen such that $z = 0$ corresponds to the incident surface. Integration over the traversed depth of the object then yields the transmitted neutron flux

$$\phi_t(z) = \phi_c \exp\left[-\int_0^z \Sigma_a(s)\,ds\right] \tag{2.11}$$

For the case of a homogeneous medium, $\Sigma_a(z)$ reduces to the constant Σ_a and therefore

$$\phi_t(z) = \phi_c \exp[-\Sigma_a z] \tag{2.12}$$

These two expressions are frequently referred to as exponential attenuation functions.

In view of our discussion above, Eqs. (2.11) and (2.12) represent the neutron flux as a function of z under conditions that neutron absorption is the most important neutron-nucleus interaction process up to coordinate z. Hence, Eq. (2.12) represents those incident neutrons which have not collided with a nucleus for a distance of z units and the expression $\phi(z)$ may be called the uncollided neutron flux.

The attenuation expression of Eq. (2.12) can be used to calculate an absorption "half-thickness" for objects of common interest. This half-thickness, to be represented by $Z_{1/2}$, is the object thickness required for the incident neutron flux to be reduced by 50% by absorption. That is, $Z_{1/2}$ is a solution of the equation

$$\Phi_c \exp[-\Sigma_a Z_{1/2}] = \frac{1}{2}\Phi_c \tag{2.13}$$

yielding the following half-thickness as a function of Σ_a for a given material

$$Z_{1/2} = \frac{\ell n(2)}{\Sigma_a} \tag{2.14}$$

Values of Σ_a can be very large for some naturally occurring metals yielding therefore a small half-thickness. For the case of gadolinium metal exposed to thermal neutrons, we obtain $Z_{1/2} \simeq 5 \times 10^{-4}$ cm while for other materials the half-thickness may be millimeters and even centimeters. This range of sensitivity suggests considerable discrimination capacity for compositional and geometrical variations of some materials.

2.4 SCATTERING AND ABSORBING OBJECT

An object sustaining both neutron scattering and neutron absorption, Fig. 2.4, is characterized by the following two macroscopic cross sections at any coordinate z:

$$\Sigma_a(z) = \sum_i \sigma_{a,i} N_i(z) \tag{2.15a}$$

$$\Sigma_s(z) = \sum_i \sigma_{s,i} N_i(z) \tag{2.15b}$$

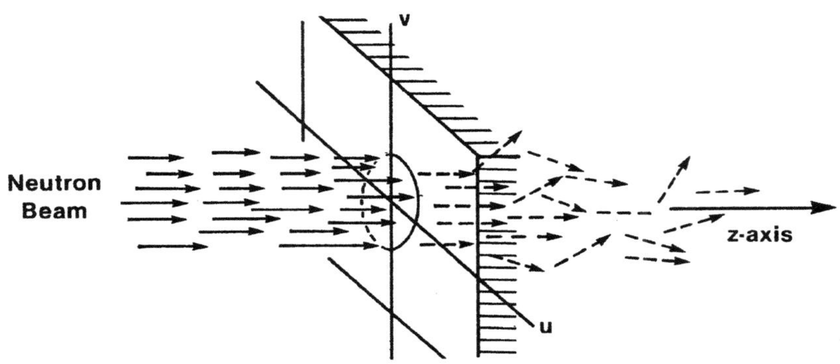

Fig. 2.4: Attenuation of a collimated neutron beam in a scattering and absorbing object.

The total macroscopic cross section at a coordinate z is given by

$$\Sigma_t(z) = \Sigma_a(z) + \Sigma_s(z) \tag{2.16}$$

and by extension of the discussion of Sec. 2.3, we may write the uncollided neutron flux

$$\Phi_{unc}(z) = \Phi_c \exp\left[- \int_o^z \Sigma_t(s)\, ds \right] \tag{2.17}$$

with the evident simplification should the medium be homogeneous.

Further thought on Fig. 2.4 and Eq. (2.17) suggests that the total neutron flux at a given coordinate z in the object consists of the uncollided component, Eq. (2.17), and of those neutrons which have undergone scattering. While a precise determination of the neutron flux at any point in the medium would require a comprehensive and detailed analysis, it is possible to combine formally the scattering and absorbed neutron flux in the following manner.

The neutron flux at any coordinate z consists of those neutrons which have not experienced a scattering or absorption interaction -- that is, the uncollided component -- and those neutrons which have experienced only scattering interactions. We therefore write the total neutron flux as

$$\Phi_{tot}(z) = \Phi_{unc}(z) + \Phi_{scat}(z) \tag{2.18}$$

where the uncollided component is given by Eq. (2.17). If we define a "build-up" function by

$$B(\Sigma_s, \Sigma_a, z) = \Phi_{tot}(z)/\Phi_{unc}(z) \tag{2.19}$$

then the scattered neutron flux component can be incorporated into the total neutron flux as

$$\Phi_{tot}(z) = \Phi_c\, B(\Sigma_s, \Sigma_a, z) \exp\left[- \int_o^z \Sigma_t(s)\, ds \right] \tag{2.20}$$

with the requirement for an evaluation or a tabulation of $B(\Sigma_s,\Sigma_a,z)$.

For those materials known to possess a very pronounced scattering cross section -- for example, hydrogenous compounds and elements such as silicon, nickel, copper, and other metals -- neutron build-up is most significant. We will show in subsequent sections that scattered neutrons can contribute substantially to image degradation.

2.5 NEUTRON CONVERTER

The neutron converter consists of a material layer adjacent to the image recorder which typically is a photographic film, Fig. 1.1. Its function is to convert the compositional and geometrical information about the object carried by the transmitted neutron beam into a film blackening radiation which is either emitted promptly, as in direct neutron radiography, or emitted according to the characteristic radioactive decay process of the neutron activated nuclei in the converter, as in indirect neutron radiography.

For our purposes here it is sufficient to take the recorder-converter arrangement as two adjacent layers on the x-y plane, as suggested in Fig. 1.1, and depicted in more detail in Fig. 2.5. In order to make our present analysis tractable, we suppose that the neutron beam is sufficiently large in area and also spatially uniform so that all physical processes are uniform in the x and y directions. Our analysis will therefore be directed only towards the elucidation of processes dependent on the z coordinate. Hence, we consider neutron absorption in the converter and the concurrent production of film blackening radiation which subsequently emerges from the converter. This secondary radiation is represented in Fig. 2.5 by ψ. Note that the neutrons impinging on the image recorder pass through it unattenuated.

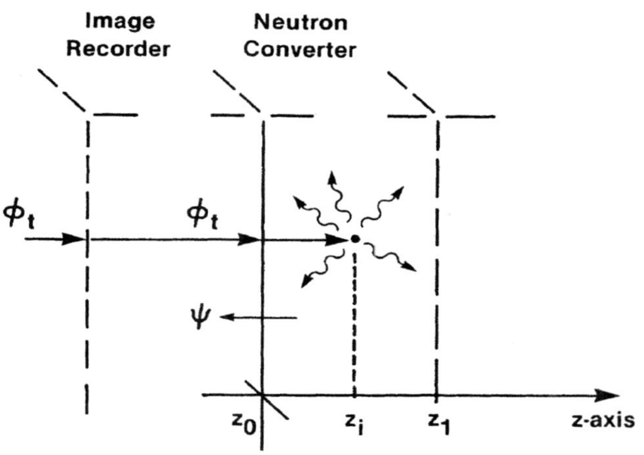

Fig. 2.5: Neutron capture in a converter and the production of film blackening radiation.

Referring to Figs. 1.4 and 2.5, we take the neutron beam transmitted through the object and incident on the converter, ϕ_t, as spatially homogeneous on the x-y plane and we assume that the converter is a highly neutron absorbing material. According to Sec. 2.3, the neutron flux at an arbitrary z-coordinate in the converter, z_i in Fig. 2.5, is given by

$$\phi(z_i) = \phi_t \exp[-\Sigma_a(z_i - z_o)] \tag{2.21}$$

Here, Σ_a is the macroscopic neutron absorption cross section for the homogeneous converter.

The neutron absorption rate per unit volume at coordinate z_i is given by

$$\begin{aligned} R_a(z_i) &= \Sigma_a \phi(z_i) \\ &= \Sigma_a \phi_t \exp[-\Sigma_a(z_i - z_o)] \end{aligned} \tag{2.22}$$

and possesses units of (reactions)/(cm^3–s). The yield rate of secondary radiation at z_i is directly proportional to the neutron absorption rate and hence may be expressed as

$$\begin{aligned} Y(z_i) &= \varepsilon_1 R_a(z_i) \\ &= \varepsilon_1 \Sigma_a \phi_t \exp[-\Sigma_a(z_i - z_o)] \end{aligned} \tag{2.23}$$

where ε_1 is a proportionality constant.

This secondary radiation may consist of γ-ray photons, alpha particles, beta particles, or internal conversion electrons, depending upon the chosen converter material. It may or may not escape from the converter but if it does, that fraction entering the film will lead to the formation of developable centers in the film (B.5). Hence, we consider the probability $P(z_i \rightarrow z_o)$ that a secondary radiation photon produced at z_i reaches z_o, Fig. 2.5. Recognizing that secondary radiation can be produced at any coordinate z_i in the converter, the total flux of secondary radiation emerging from the converter at z_o follows by integration of $Y(z_i)$, Eq. (2.23), multiplied by its probability of escape $P(z_i \rightarrow z_o)$, over the entire converter thickness. That is, the secondary radiation flux entering the film is given by

$$\begin{aligned} \psi &= \int_{z_o}^{z_1} Y(z_i)\, P(z_i \rightarrow z_o)\, d z_i \\ &= \varepsilon_1 \Sigma_a \phi_t \int_{z_o}^{z_1} \exp[-\Sigma_a(z_i - z_o)]\, P(z_i \rightarrow z_o)\, d z_i \end{aligned} \tag{2.24}$$

An explicit expression for $P(z_i \rightarrow z_o)$ is now required. Secondary radiation is invariably produced isotropically and is generally attenuated exponentially in materials of common interest. Hence, we use,

$$P(z_i \rightarrow z_o) = \exp[-\mu(z_i - z_o)] \tag{2.25}$$

where μ is an effective attenuation parameter for the secondary radiation in the converter; that is, μ accounts for emission over all angles and is therefore not to be equated with the linear attenuation parameter of a specific ray of secondary radiation. Substitution of Eq. (2.25) into Eq. (2.24) leads to a particularly compact expression and accounts for all forms of radiation emission and their directions:

$$\psi = \varepsilon_1 \Sigma_a \Phi_t \int_{z_o}^{z_1} \exp[-\Sigma_a(z_i - z_o)] \exp[-\mu(z_i - z_o)] d z_i$$

$$= \frac{\varepsilon_1 \Sigma_a \Phi_t}{(\Sigma_a + \mu)} \{1 - \exp[-(\Sigma_a + \mu)(z_1 - z_o)]\} \tag{2.26}$$

Thus, the specific role of various parameters in effecting the magnitude of the converter response radiation flux is explicitly indicated. We will consider an experimental test of this relationship following a discussion of the coupling between a converter and film.

2.6 CONVERTER-RECORDER COUPLING

The image formation process begins with the secondary radiation flux, ψ, produced by the converter, being absorbed in the recorder. Restricting our discussion again to film recorders, the common process assumed (B.5) is that the secondary radiation is absorbed by silver halide grains in the film emulsion during the exposure time, τ. Through the action of silver reduction these exposed grains collectively form a latent image. The development process serves catalytically to extend the reduction of silver so that far more silver is precipitated by exposed grains than by unexposed grains. This precipitated silver results in an optical blackness, or optical density patterns forming the final image. The degree of blackening, that is the quantified optical density, represents the result of a neutron radiograph measurement. Two considerations are important in this process: one is the total exposure -- or total fluence -- of secondary radiation from the converter onto the recording film and the other is the film response to this exposure.

The exposure of the film at an arbitrary point (x,y) on the image plane, Figs. 1.4 and 2.5, is given by the integration of the film blackening radiation $\psi(x,y)$ over time. For the general case we consider $\psi(x,y)$ as time dependent, writing $\psi(x,y) \rightarrow \psi(x,y,t)$, and so define the film exposure at the (x,y) coordinate by E(x,y) with

$$E(x,y) = \int_0^{\tau_f} \psi(x,y,t)\,dt \tag{2.27}$$

where τ_f is the time interval during which secondary radiation from the converter impinges on the film recorder.

The case of direct neutron radiography is now easily treated. The film exposure time, τ_f, is equal to the exposure time, τ_n, that the object is exposed to the neutron beam and $\psi(x,y,t)$ is temporally constant, denoted by $\psi(x, y)$. We therefore have

$$E(x,y) = \psi\,(x,y)\,\tau_n \tag{2.28}$$

Film exposure associated with indirect neutron radiography incorporates the radioactive decay property of the nuclei in the converter which have been activated by the capture of neutrons from the transmitted neutron beam. The time dependence of the film blackening secondary radiation flux follows from the exponential nature of radioactive decay and we write for this case

$$E(x,y) = \int_{t_1}^{t_2} \psi(x,y,t)\,dt$$

$$= \int_{t_1}^{t_2} \psi(x,y,t_1)\,e^{-\lambda t}\,dt \tag{2.29}$$

Here, λ is the decay constant of the activated isotope, $\psi(x,y,t_1)$ is the magnitude of the film blackening radiation at the beginning of the film-converter contact time, t_1, and t_2 is the time when the film is removed from the converter. Hence, for indirect neutron radiography, the film exposure time, $\tau_f = t_2 - t_1$, is not necessarily the neutron exposure time, τ_n.

Integration of Eq. (2.29) is readily possible and the expression for the indirect case of neutron radiography is given by

$$E(x,y) = \frac{\psi(x,y,t_1)}{\lambda}\left(e^{-\lambda t_1} - e^{-\lambda t_2}\right) \tag{2.30}$$

Figure 2.6 provides a graphical depiction of the exposure relations for the direct and indirect methods of neutron radiography.

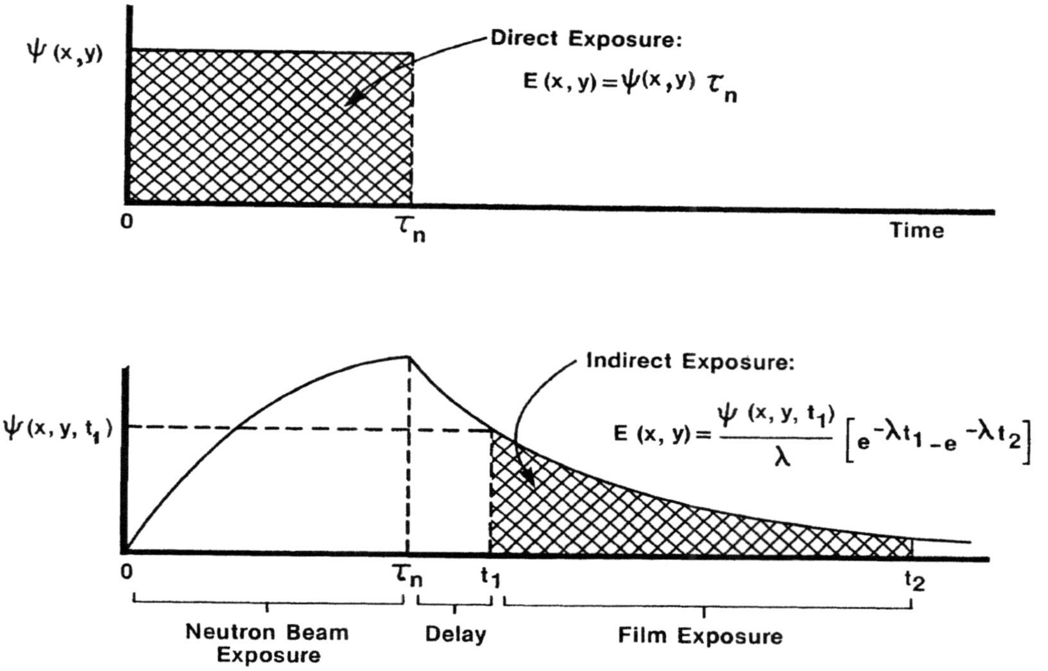

Fig. 2.6: Graphical depiction showing film exposure for direct and indirect neutron radiography.

While film exposure, E(x,y), is determined by considerations of (i) neutron beam exposure time, τ_n, (ii) the transmitted neutron beam, $\phi_t(x,y)$, and (iii) physical characteristics of the converter, the eventual optical density induced is further a function of the type of radiation generated by the converter and the development process. A primary consideration then is the direct relationship between the optical density at (x,y) in the film, D(x,y), and the exposure, E(x,y), generated by the converter.

Since the exposure is a time-integrated flux, it possess units of cm^{-2}. That is, E(x,y) represents the cumulative effect of the number of radiation "bullets" leaving the converter, per unit area, and becoming "hits" in the film. By a photochemical process (B.5) these "hits" form an optical blackening, the magnitude of which yields the optical density, D(x,y). This optical density is measured at a point (x, y) by focussing a light beam onto the film and taking the logarithm of the inverse transmitted light fraction at that point. The relation is

$$D(x,y) = \log\left[\frac{I_o(x,y)}{I_f(x,y)}\right] \tag{2.31}$$

where $I_o(x,y)$ is the total light intensity incident on the film and $I_f(x,y)$ is the transmitted light intensity.

The relationship between the optical density, D(x,y), and the exposure, E(x,y), depends upon the kind of film, the type of radiation emitted by the converter and the process of film development. For a specified development process and given converter, the exposure-density relationship may be highly nonlinear, as indicated in Fig. 2.7. The nonlinearity between the optical density and the film exposure means that a direct proportionality exists only over a finite exposure interval. When a precise relationship between the material and/or geometrical properties of the object being radiographed and the film blackening radiation can be established, the direct translation of this relationship to an optical density variation requires either that the exposure-density variation occurs in the linear range or that it be specifically determined. By consideration of the exposure-density variation, optical density can thus be related to object or converter properties. Figure 2.8 shows the satisfactory agreement obtainable between theory, Eq. (2.26), and experiment.

For a given neutron converter, it may be convenient to establish a D-vs-ϕ_t variation since, for a uniform beam, ϕ_t is proportional to, and more easily measurable than, ψ. We will see in Sec. 3.1 that a log-linear exposure-density variation may also be useful.

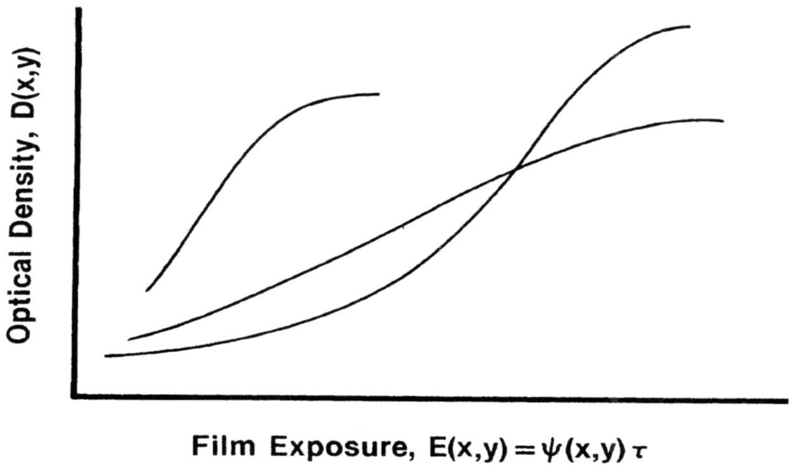

Fig. 2.7: Sample relationships between film exposure and optical density.

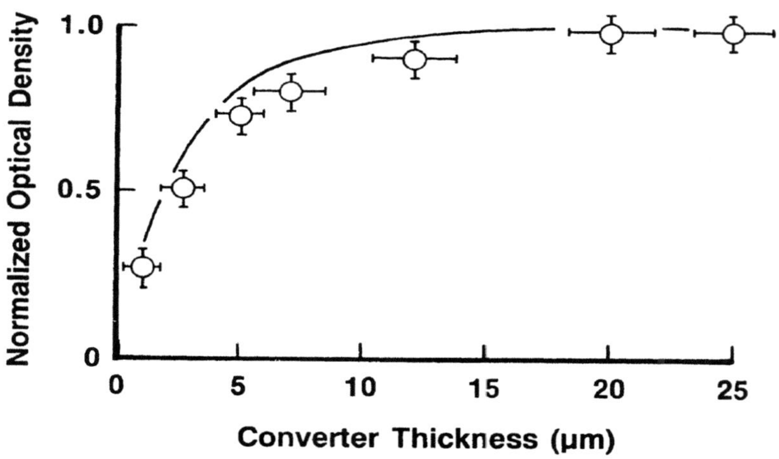

Fig. 2.8: Comparison between measured optical density (open circle) and that based on Eq. (2.26) (solid line). Adapted from Ref. A.34.

While a general discussion of the exposure-density variation is beyond the scope of this section, several meaningful imaging indices, based on this variation and with their roots in X-ray radiography, are frequently employed (B.8). These indices are illustrated in Fig. 2.9 for a particular exposure-density variation in which density is plotted against log exposure. Briefly, they are defined as follows: Firstly, there is the background density, D_b, arising from the development of unexposed silver halide grains, referred to as fog, and from the opacity of the film base itself, referred to as the base density. Secondly, the "film latitude" describes the range of exposures over which the curve has a constant slope. Thirdly, the "speed" of a film, not shown on Fig. 2.9, relates to the exposure required to produce an optical density of one above the background density, thereby describing how "fast" this specific density can be achieved with the film. It is thus defined as the reciprocal of the required exposure. Finally, the film "gamma", γ, is a useful parameter since it describes the contrast magnification properties of the film. When gamma exceeds one, the film exaggerates the actual subject contrast.

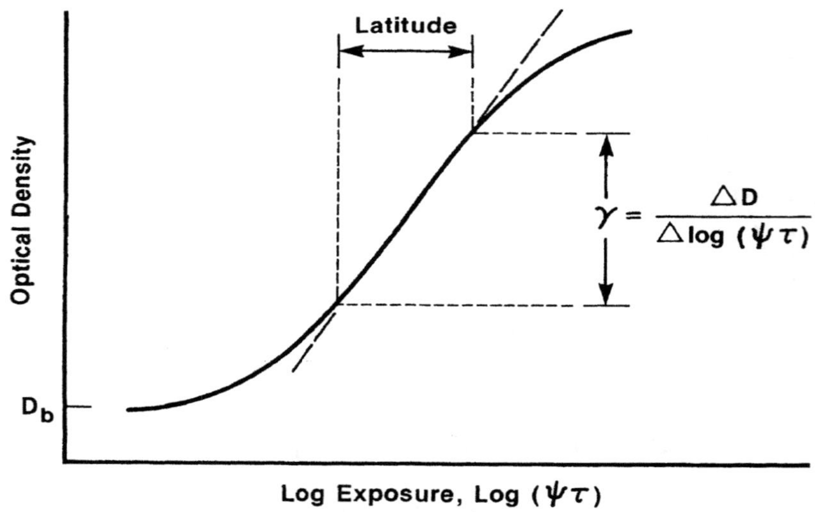

Fig. 2.9: Exposure-density variation illustrating commonly employed imaging indices.

2.7 TRACK-ETCH IMAGING

Track-etch neutron radiography is based on the use of a plastic recorder adjacent to the neutron converter; no film is necessary. The converter is chosen so that neutron absorption leads to a massive reaction product which forms a microscopic track in the recorder made visible by subsequent chemical etching, Fig. 2.10.

Common choices for the converter are compounds loaded with ^{10}B and/or ^{6}Li. Both possess a significant neutron cross section for the following reactions:

$$^{10}B + n \rightarrow \alpha + {}^{7}Li \tag{2.32a}$$

$$^{6}Li + n \rightarrow \alpha + {}^{3}H \tag{2.32b}$$

Table 2.1 lists some parameters for these reactions.

A common plastic recorder is cellulose nitrate. It may be used on both sides of the converter and, if suitably superimposed after etching, may provide the usual effect of both reaction products of Eq. (2.32).

Table 2.1

Parameters associated with boron–10 and lithium-6 reactions.

	$^{10}B(n,\alpha)^{7}Li$	$^{6}Li(n,\alpha)^{3}H$
Thermal microscopic cross section (b)	3840	945
Reaction Q-value (MeV)	2.31	4.78
Reaction product energies (MeV)	$E_\alpha = 1.47$	$E_\alpha = 2.04$
	$E_{Li} = 0.84$	$E_H = 2.74$

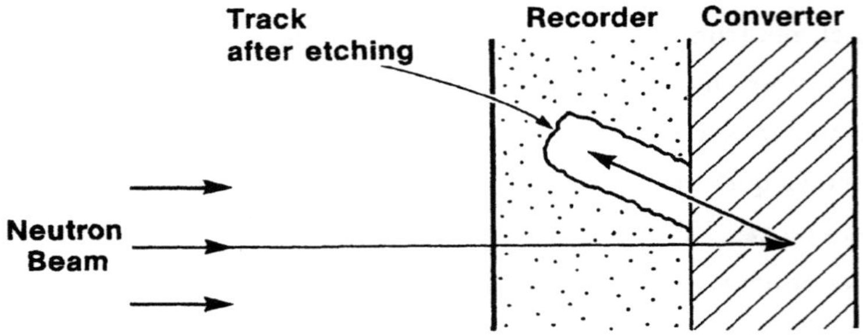

Fig. 2.10: Track-etch neutron radiography.

III. Object-Image Relations

A radiographic image conveys information about specific compositional and geometrical properties of the object. Our aim here is to identify and characterize some of these essential relations which also are to be used in subsequent chapters.

3.1 COMPOSITIONAL AND GEOMETRICAL EFFECTS

As an introductory case, we consider an object of uniform thickness but of varying composition. The compositional variations of relevance in neutron radiography are expressed in terms of the neutron absorption cross section, Σ_a, associated with the local properties of an object. Compositional variations characterized by the neutron scattering cross section, Σ_s, are also of importance but are deferred for now because this may obscure some fundamental imaging aspects we wish to emphasize. A case of interest is depicted in Fig. 3.1a; ϕ_c is the uniform neutron flux emerging from the collimator and incident on the neutron absorbing object while $\phi_{t,1}$ and $\phi_{t,2}$ are the transmitted uncollided neutron flux associated with each material region and which subsequently are incident on the neutron converter.

Following the analysis of Sec. 2.3, the transmitted neutron flux is determined by the isotopic composition of the material through which the beam passes. For objects of identical thickness but different composition, Fig. 3.1a, we write for the transmitted neutron flux

$$\phi_{t,1} = \phi_c \exp[-\Sigma_{a,1} z_o] \tag{3.1a}$$

$$\phi_{t,2} = \phi_c \exp[-\Sigma_{a,2} z_o] \tag{3.1b}$$

With each isotopically distinct material domain, we can therefore associate a converter response, ψ_1 and ψ_2, of the form suggested by Eq. (2.26), and which subsequently leads to an optical density, Sec. 2.6. That is, we have the general sequential relation

$$\phi_t \rightarrow \psi \rightarrow D \tag{3.2}$$

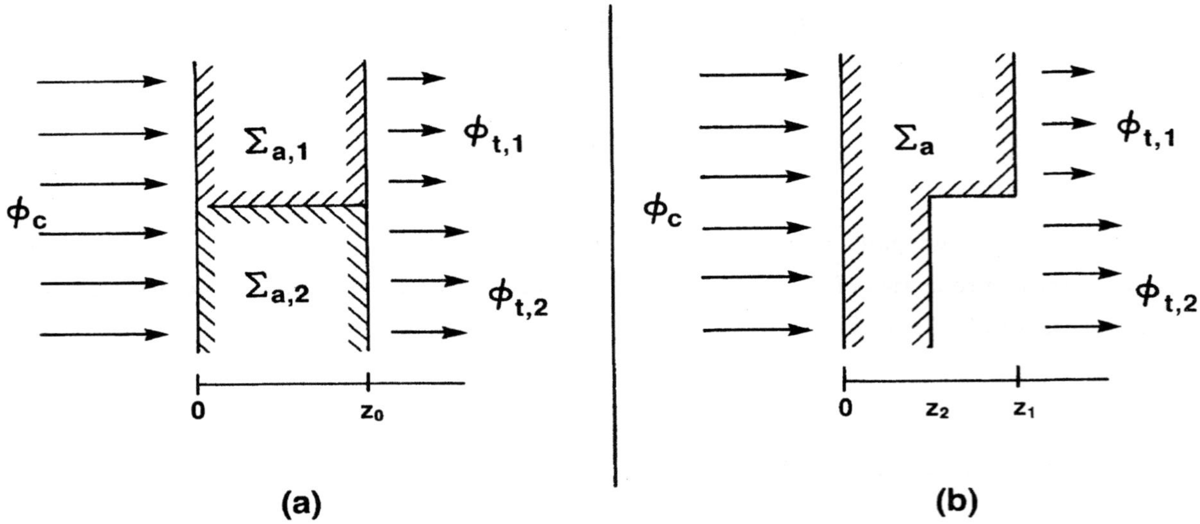

Fig. 3.1: Neutron beam penetrating an absorbing object of (a) uniform thickness but varying composition, and (b) uniform composition but varying thickness.

where D is the experimentally measured optical density on the film. Now ψ is directly related to ϕ_t, Sec. 2.5, and D is related to ψ through the exposure-density variation, Sec. 2.6, and generally given by

$$D = f(\psi\tau) \tag{3.3}$$

where τ is the exposure time . The optical densities corresponding to Eqs. (3.1) are therefore

$$D_1 = f(A \exp[-\Sigma_{a,1} z_o]) \tag{3.4a}$$

$$D_2 = f(A \exp[-\Sigma_{a,2} z_o]) \tag{3.4b}$$

where the constant A is the product of ϕ_c, τ and the proportionality constant between ψ and ϕ_t, Eq. (2.26).

As discussed in the preceding chapter, while the relationship between the transmitted neutron flux, ϕ_t, and the converter response flux, ψ, can be established, e.g., Eq. (2.26), and further that the exposure time τ can easily be determined, it is the potential nonlinear transformation from exposure to optical density that constitutes an annoying, though most important, connection. This can be illustrated by the following.

Consider the transmitted neutron flux $\phi_{t,\,1}$ and $\phi_{t,\,2}$ for either of the two cases illustrated in Fig. 3.1 at points far from the interface domain. The associated converter response flux are therefore taken as ψ_1 and ψ_2, respectively, and hence we arrive at the film exposures

$$E_1 = \psi_1\,\tau \tag{3.5a}$$

$$E_2 = \psi_2\,\tau \tag{3.5b}$$

A typical transformation of these two exposures into optical densities for two different film types is suggested in Fig. 3.2. The two important results are that both the magnitude of optical density and the optical density difference are profoundly affected by the film properties. It is experience which invariably determines the film type, as well as the developmental process and the exposure time, in accordance with certain radiographic objectives.

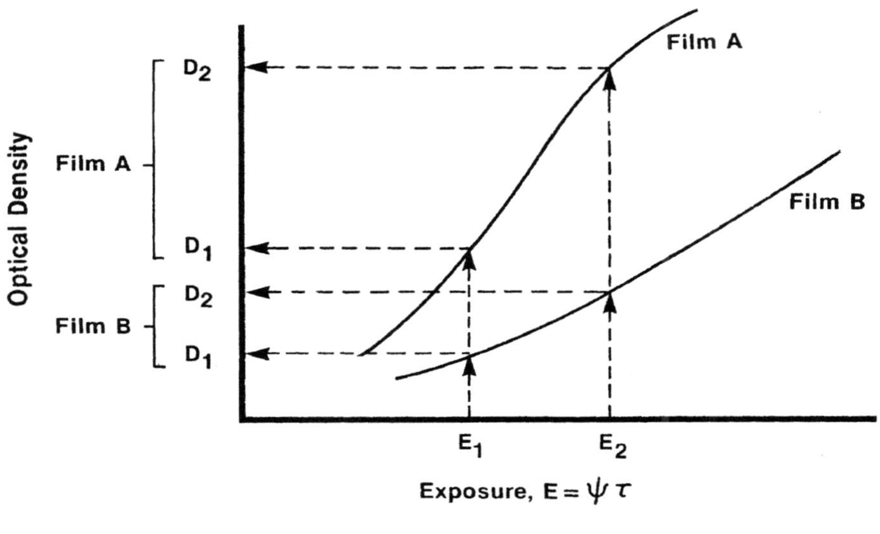

Fig. 3.2: Effect of different films on optical density for the case of identical film exposure.

It is important to point out here the general utility of an exposure in either the linear or log-linear range of the exposure-density variation. For the linear range, densities are given by

$$D = a + b\psi\tau \tag{3.6}$$

for suitable constants a and b. The "contrast", or density ratio, generated by the two exposures is therefore

$$\frac{D_2 - a}{D_1 - a} = \frac{\psi_2}{\psi_1} = \begin{cases} \exp[(\Sigma_{a,1} - \Sigma_{a,2})z_o], & \text{for Fig. 3.1a} \\ \exp[\Sigma_a(z_1 - z_2)], & \text{for Fig. 3.1b} \end{cases} \tag{3.7}$$

For the simple geometries of Fig. 3.1, a quantitative discrimination between regions is thus possible.

Likewise, for the log-linear range, the "contrast" or density difference, generated by the two exposures is

$$D_2 - D_1 = \begin{cases} \gamma[(\Sigma_{a,1} - \Sigma_{a,2})z_o], & \text{for Fig. 3.1a} \\ \gamma[\Sigma_a(z_1 - z_2)], & \text{for Fig. 3.1b} \end{cases} \tag{3.8}$$

where γ is the film gamma.

3.2 IMAGE DEGRADATION

In our analysis of the optical density associated with regional compositional and geometrical variations, we have been careful in avoiding a discussion of the transition region. As depicted in Fig. 3.3 for the region in which either the composition or the geometry of the object changes significantly, an optical density variation occurs which asymptotically approaches the two dominant trends. The optical density variation about a material edge is frequently called "edge blurring" or "edge unsharpness", and represents a special form of image degradation.

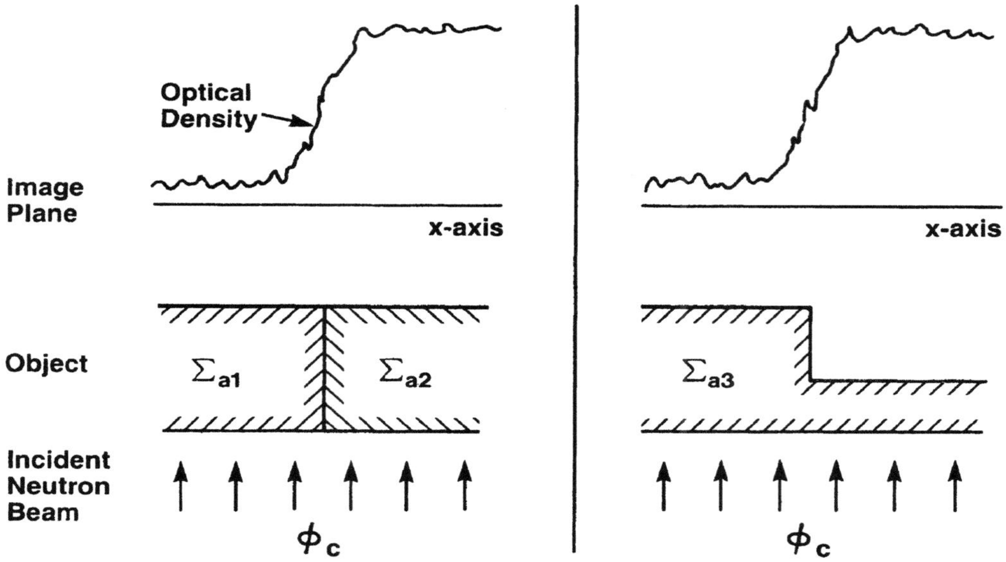

Fig. 3.3: Optical density variation due to differences in material composition and in material thickness.

Image degradation in general, and edge unsharpness in particular, can be influenced by a number of system components of which the following are specific examples:

a) Converter unsharpness: the capture of a neutron in a converter leads to the isotropic emission of secondary radiation which may lead to the formation of a developable center in the film at some orthogonal distance from the incident neutron path.

b) Object scattering degradation: neutrons scattered in the object lose their collimation leading to a reduction in contrast and possibly local image artifacts.

c) Geometric unsharpness: an increasing beam divergence will tend to geometrically increase the unsharpness.

d) Motion unsharpness: object motion during an exposure contributes to image blurring.

e) System noise: ideal optical density variations are contaminated by film-grain noise as well as measurement induced electronic and photoelectronic noise.

Each of these image degradation contributions will be considered at some point in this text. However, in order to establish a conceptual and mathematical basis for our subsequent analysis we consider first converter unsharpness associated with a material edge of the type suggested in Fig. 3.3. Indeed, we will simplify the analysis by considering only the unsharpness associated with a totally absorbing knife-edge object and a highly collimated neutron beam.

3.3 LINE-SPREAD UNSHARPNESS

The sequence of neutron capture in the converter and the ensuing formation of developable centers in the film is a distinguishing feature in neutron radiography. Its relationship to image degradation arises from secondary radiation contributing to optical density formation at some orthogonal distance from the initiating neutron path. We examine first the image degradation effect of this process taking care to highlight the underlying conceptual-mathematical methodology.

Consider the situation shown in Fig. 3.4. Here, a collimated beam of neutrons is totally absorbed in an object except for an "infinitesimally" narrow line source orthogonal to the x-axis. Neutron capture is taken to occur in the converter at z_1 and on the line x_0 with secondary radiation emitted isotropically. Our objective is to determine the extent to which this radiation reaches the film at any x-coordinate; that is, in terms of our previous discussion, we need to determine the converter response flux, ψ.

The geometry of Fig. 3.4 has been chosen in order to facilitate an analytical description. An important feature is that the narrow slit of neutrons passing through the object and the film is uniform in the y-direction so that no useful information is contained as a function of y. Further, we have compressed the converter and the film into "thin" planes at z_1 and z_0 respectively eliminating therefore all spatial integrations in the z-direction. Thus, with the film-converter separation given by a constant $(z_1 - z_0)$, our analysis needs only to incorporate a functional dependence of ψ on the x variable at the image plane, i.e. $\psi(x)$.

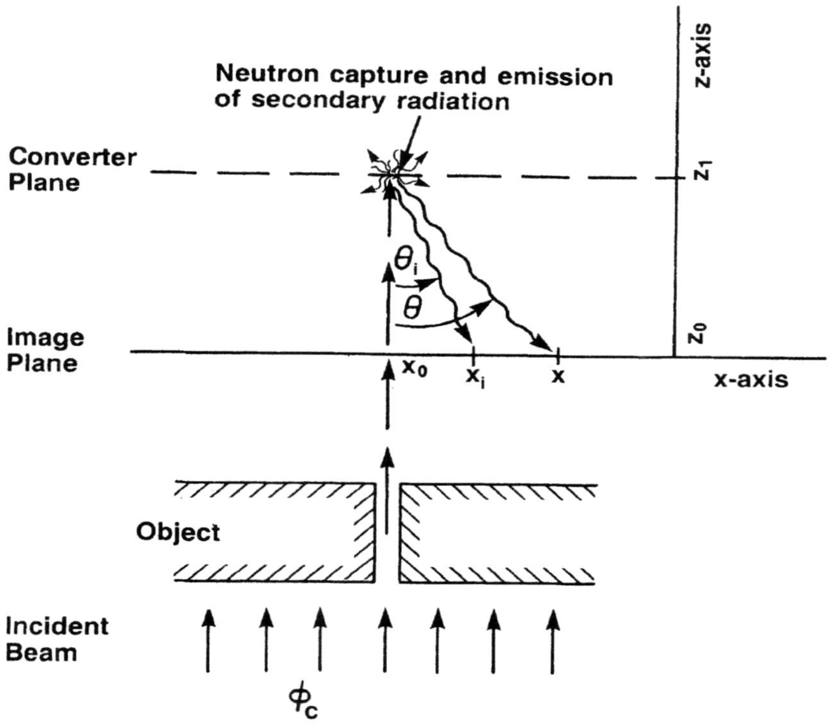

Fig. 3.4: Geometrical representation of converter emitting isotropic secondary radiation onto the image plane.

Consider first the upper part of the x-axis for all $x \geq x_o$, Fig. 3.4, and introduce the function $G(x, x_o)$ as the probability that the secondary radiation whose origin is at the line through (z_1, x_o) will contribute to the response flux $\psi(x_i)$ in the interval $x_i \in (x_o, x)$. This can be written formally as

$$G(x, x_o) = \Pr[x_o \leq x_i \leq x] \qquad (3.9)$$

where Pr[ ] represents the probability of the event described in the brackets [ ]. By our definition of the angle θ_i, Fig. 3.4, we write equivalently

$$G(x, x_o) = \Pr\left[\tan\theta_i \le \left(\frac{x - x_o}{z_1 - z_o}\right)\right]$$

$$= \Pr\left[\theta_i \le \tan^{-1}\left(\frac{x - x_o}{z_1 - z_o}\right)\right], \quad x \ge x_o \tag{3.10}$$

Note that θ_i is measured from the vertical line, as shown in Fig. 3.4, in the plane of the diagram. This is not intended to convey that all secondary radiation is emitted in this plane, but rather that the distribution of secondary radiation can be calculated with a knowledge of only one of the two polar emission angles, that being θ.

For isotropic emission of the secondary radiation it is evident that

$$\Pr[\theta_i \le \theta] = \frac{2\theta}{\pi}, \quad \theta \in \left(0, \frac{\pi}{2}\right) \tag{3.11}$$

so that, explicitly for $G(x, x_o)$, Eq. (3.10) becomes

$$G(x, x_o) = \frac{2}{\pi}\tan^{-1}\left(\frac{x - x_o}{z_1 - z_o}\right), \quad x \ge x_o \tag{3.12}$$

By the definition of Eq. (3.9), this function $G(x, x_o)$ is the cumulative probability function for the converter response up to x with the secondary radiation source at the line x_o on the z_1 plane. However, it is commonly found more useful to use a probability density function rather than a cumulative probability function. Introducing $g(x, x_o)$ as the probability density function corresponding to $G(x\ x_o)$, we use the well known relationship between such functions

$$g(x, x_o) = \frac{d}{dx}[G(x, x_o)] \tag{3.13}$$

to obtain specifically

$$g(x, x_o) = \frac{2}{\pi(z_1 - z_o)\left[1 + \left(\frac{x - x_o}{z_1 - z_o}\right)^2\right]}, \tag{3.14}$$

Symmetry of $g(x, x_o)$ about $x = x_o$ suggests that the restriction for $x \ge x_o$ can be removed. The resulting secondary radiation probability density function can be written as

$$g(x, x_o) = \left[\frac{\lambda/\pi}{1 + \lambda^2(x - x_o)^2}\right] \tag{3.15}$$

where

$$\lambda = (z_1 - z_o)^{-1} \tag{3.16}$$

is taken as a system parameter.

The above probability density function is of such general importance that we will refer to it as the line-spread function for which we introduce the symbol $L(x,x_o)$:

$$L(x, x_o) = \frac{\lambda/\pi}{1 + \lambda^2 (x - x_o)^2} \tag{3.17}$$

Note that x_o is an arbitrary coordinate location for the infinitesimally narrow slit. Choosing $x_o = 0$ leads to a particularly simple form

$$L(x) = \frac{\lambda/\pi}{1 + \lambda^2 x^2} \tag{3.18}$$

Both the one and two variable notations, $L(x)$ and $L(x, x_o)$, are common and used interchangeably or as dictated by the notational context. Figure 3.5 depicts this function for three values of the resolution parameter λ.

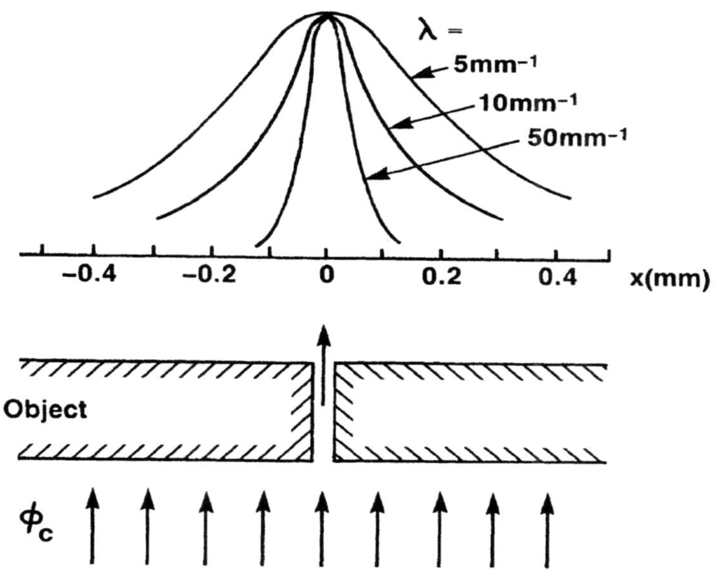

Fig. 3.5: Line-spread function, $L(x)$, for different values of the resolution parameter, λ.

To interpret the line-spread function physically, we note that if it were possible to obtain a completely absorbing object with an infinitesimally narrow slit, then the converter response, $\psi(x, x_0)$, would be directly proportional to $L(x, x_0)$. If the film possessed and was used in a linear exposure-density range, then the optical density would also be expected to possess a variation as suggested in Fig. 3.5. Whether Eq. (3.18) is indeed the appropriate function to use depends on whether or not the important and dominant processes are adequately represented in its derivation. When several other image degradation sources, each of which generates a line-spread function, are combined with converter unsharpness, central limit considerations dictate that the overall line-spread function may approach a Gaussian form:

$$L(x) = \frac{\alpha}{\sqrt{\pi}} \exp[-\alpha^2 x^2] \tag{3.19}$$

Note that Eq. (3.19) is a comprehensive line-spread function while Eq. (3.18) is of particular relevance in high resolution neutron radiography for which converter unsharpness is the dominant image degradation source. A comparison of these two functions is provided in Sec. 9.2. While other functional forms are of course possible, we will here use only the line-spread function in Eq. (3.18) and emphasize that an experimental test is required to confirm or deny a particular choice for a line-spread function.

3.4 FULL-WIDTH-AT-HALF-MAXIMUM

The availability of an appropriate line-spread function is invariably of considerable conceptual and analytical utility. For example, image sharpness and image system resolution capacity can be related to the relative "narrowness" or "width" of the line-spread function. A narrow line-spread function is a much sought goal of a radiographic imaging system.

The narrowness or spreading of the line-spread function in Eq. (3.18) is completely determined by the parameter λ; indeed, the line-spread function width will be shown to be inversely proportional to this parameter. This resolution property can be illustrated by determining the full-width-at-half-maximum of $L(x)$, as depicted in Fig. 3.6.

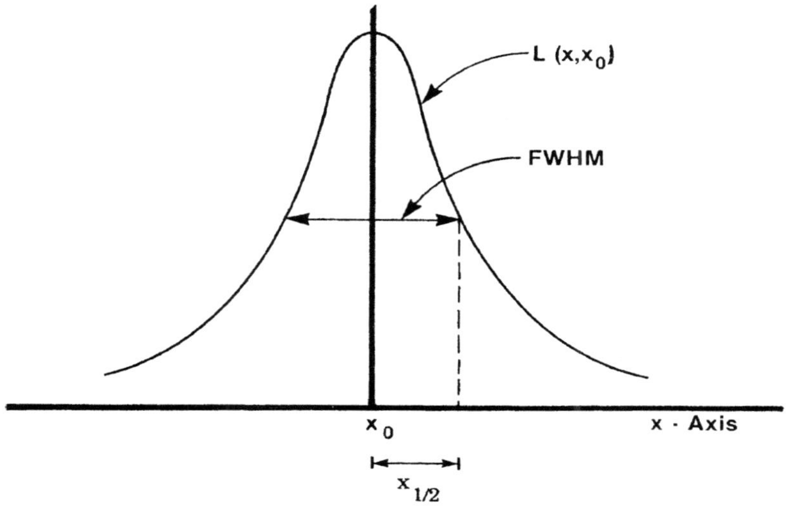

Fig. 3.6: Full-width-at-half-maximum for the line-spread function $L(x,x_0)$.

For the line spread function of Eq. (3.22) the x-coordinate corresponding to one-half of the maximum of L(x) -- to be represented by $x_{1/2}$ -- is a solution of

$$\frac{1}{2}(\lambda/\pi) = \frac{\lambda/\pi}{1+\lambda^2 x_{1/2}^2} \tag{3.20}$$

This yields explicitly

$$x_{1/2} = \frac{1}{\lambda} \tag{3.21}$$

and the full-width-at-half-maximum, FWHM, is double this magnitude

$$\mathrm{FWHM} = \frac{2}{\lambda} \tag{3.22}$$

Thus, the larger λ, the smaller the FWHM and the narrower the line-spread function. High resolution neutron imaging would therefore be characterized by large values of this resolution parameter.

3.5 EDGE-SPREAD UNSHARPNESS

While a line-spread function is of fundamental radiographic interest, it is difficult to obtain experimentally since, in practice, it is impossible to obtain an infinitesimally narrow line source of neutrons. However, if one part of the object of Fig. 3.5 is removed, then an indirect determination of the line-spread function can be undertaken.

In Fig. 3.7, we illustrate a knife-edge object obtained by the elimination of one part of the completely absorbing object of Fig. 3.5. This same figure also depicts two line-spread functions associated with two arbitrarily selected isolated infinitesimally narrow line sources of neutrons. Intuition suggests that the total converter and film responses leading to the optical density variation for this ideal knife-edge object can be constructed as the limiting sum of line responses to closely spaced line sources over the $x \geq 0$ domain of Fig. 3.7. This process of linear superposition is exactly described by integration of the line-spread function, to yield the converter response

$$S(x) = N \int_0^\infty L(x, x_0)\, dx_0 \tag{3.23}$$

Here, S(x) will be called the edge-spread function (A.35) because it relates to the blurring of image information about the edge of an ideal absorbing object. The symbol N represents an arbitrary normalization parameter.

For a line-spread function of interest to neutron radiography, derived in Sec. 3.3 and expressed by Eq. (3.17), an explicit expression for the edge-spread function, S(x), is therefore given by

$$S(x) = N \int_0^\infty \left(\frac{\lambda/\pi}{1 + \lambda^2 (x - x_0)^2} \right) dx_0 \tag{3.24}$$

This integration yields

$$S(x) = N \left[\frac{1}{2} + \frac{1}{\pi} \tan^{-1}(\lambda x) \right] \tag{3.25}$$

We depict this function in Fig. 3.8 for the same three values of the resolution parameter λ used in Fig. 3.5. By setting $N = 1$, a convenient normalization is obtained whereby the edge-spread function lies between zero and one.

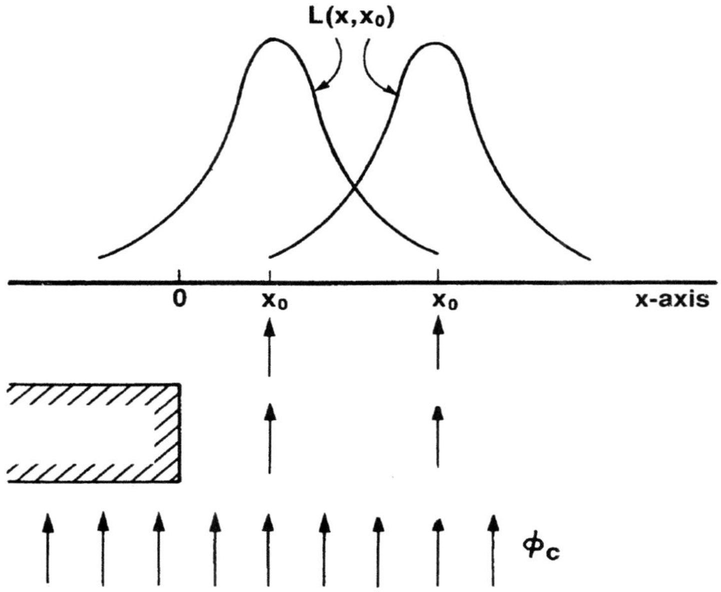

Fig. 3.7: Image plane response as a linear superposition of line-spread functions.

For an ideal radiographic system with no image degradation, $\lambda \to \infty$, resulting in a step edge-spread function and Dirac delta (unit impulse, B.6) line-spread function.

3.6 EXTENSION OF THE LINE-SPREAD FUNCTION

The line-spread and edge-spread functions of Secs. 3.3 and 3.5 are mathematical descriptions of the converter response radiation which impinges on a radiographic film, and are associated with line and half-plane collimated neutron sources. While these are of fundamental descriptive utility, it is ultimately the corresponding optical density variations that are of interest to the radiographer. The conversion of exposures to optical densities is analytically represented by the transformation of exposures through the exposure-density variation, Sec. 2.6. Some additional comments are now appropriate.

Consider the ideal neutron absorbing knife-edge object of Fig. 3.8. An actual microdensitometer scan of the knife-edge neutron radiograph would yield three superimposed optical density components, as shown in Fig. 3.9:

$$D(x) = D_{object}(x) + D_{noise} + D_{background} \tag{3.26}$$

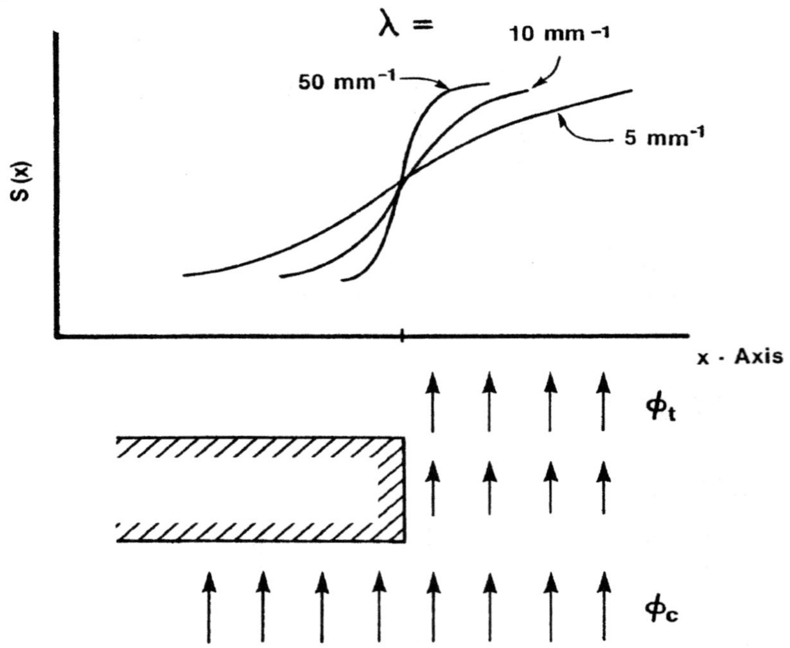

Fig. 3.8: Edge-spread function, S(x), for different values of the resolution parameter, λ.

For algebraic and notational convenience, we will simply write

$$D(x) = D_o(x) + D_n + D_b \tag{3.27}$$

The background component, D_b, should be uniform and independent of the object's dimensions or edge location. The system noise, D_n, is attributable to statistical variations in the film-grain distribution and electronic signals as well as short-range chemical processing effects. This component is also taken to be independent of the object (to be further discussed in Ch. IX). Thus, only $D_o(x)$ is specifically associated with the object.

In practice, D_b is identified as the optical density on a portion of the film not directly exposed to the neutron beam, while D_n can removed by either mathematical filtering, (Sec. 9.8), or more simply, by visual "smoothing". The smoothed optical density component, $D_o(x)$, associated with the object can thus be isolated.

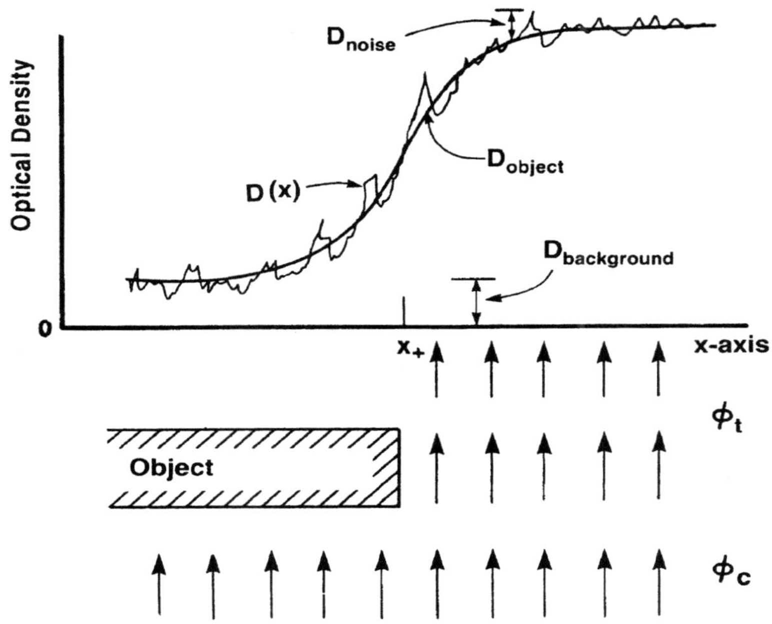

Fig. 3.9: Optical density components associated with a neutron absorbing knife-edge object.

Now, the exposure-density variation governs the transformation of converter responses to optical densities, enabling the analytical derivation of $D_0(x)$ from S(x). If, in particular, this exposure-density variation is in a linear range, then a conveniently normalized optical density variation across the image of a neutron absorbing knife-edge object, for a collimated neutron beam, is given by

$$D(x) = \frac{1}{2} + \frac{1}{\pi} \tan^{-1}(\lambda x) \tag{3.28}$$

The resolution parameter, λ, in this equation is obtained by a fit to the experimentally obtained D(x) by, say, the method of least squares. For several applications of image interpretation, however, we will see that a determination of λ is not always necessary.

Any degree of confidence in the above theoretical model is clearly dependent upon experimental test. While, as we indicated, it is essentially impossible to directly confirm experimentally the functional form of the line-spread function, an experimental test of the edge-spread function is easily possible. We illustrate such a comparison in Fig. 3.10, which suggests that the edge-spread function in Eq. (3.25) can be sufficiently accurate.

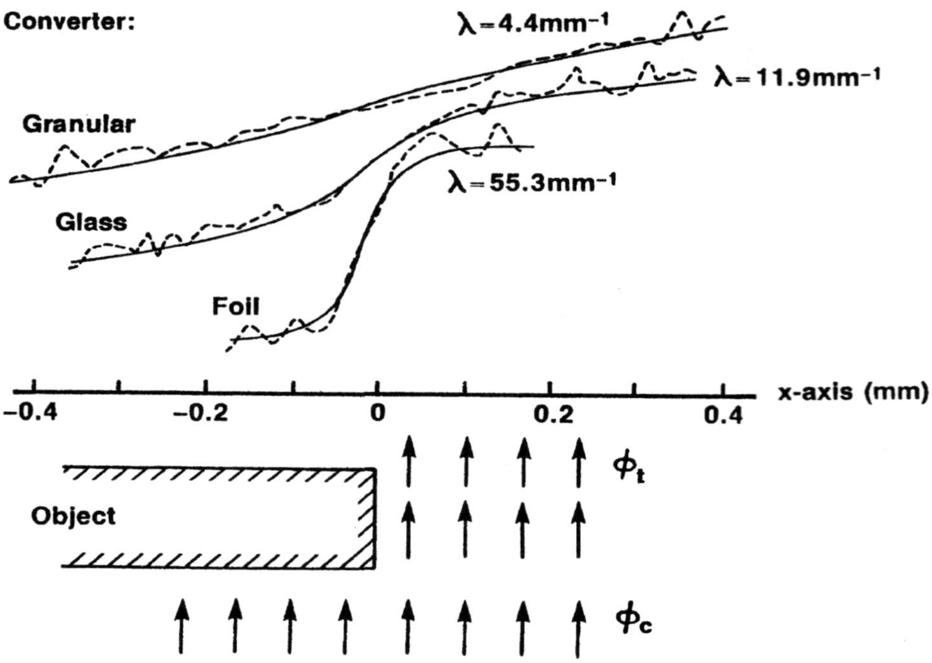

Fig. 3.10: Comparison between experimentally measured edge-spread functions and those obtained using Eq. (3.28) with λ fitted. (A.43, p.82); each function is normalized to the interval from zero to unity on the optical density coordinate.

The approach used in the preceding section, particularly that leading to Eq. (3.28), masks a very powerful tool of mathematical analysis. We illustrate this by rewriting Eq. (3.23) for the case depicted in Fig. 3.7 using the notation

$$S(x) = N \int_{-\infty}^{\infty} \phi_t(x_0)\, L(x, x_0)\, dx_0 , \tag{3.29}$$

where the transmitted neutron flux is given by

$$\phi_t(x_0) = \begin{cases} 0, & x_0 \leq 0 \\ 1, & x_0 > 0 \end{cases} \tag{3.30}$$

Here, $\phi_t(x_o)$ represents the neutron beam actually impinging on the converter for the case of a perfectly absorbing object with its edge located at $x_o = 0$. Hence, $\phi_t(x_o)$ for this case is a Heaviside step function (B.6).

The significant feature of Eq. (3.29), however, rests on its generality for an arbitrary function $\phi_t(x_o)$. The reader may recognize that Eq. (3.29) describes the process of "convolving" an input function, $\phi_t(x_o)$, with a response function, $L(x, x_o)$. This process is referred to as "convolution" (B.6) because, in general, the convolution of an input function with a positive response function tends to smoothen variations in the input function. In this instance, the input step function is convolved with a line-spread response function to produce a smooth converter response function.

We have thus established the general expression for the converter radiation response, $\psi(x)$, which ultimately relates to the optical density of the radiographic film or other recorder response, as a function of the transmitted neutron flux, $\phi_t(x_o)$, and the intrinsic converter unsharpness, $L(x,x_o)$, in complete one-dimensional generality as

$$\begin{aligned}\psi(x) &= N \int_{-\infty}^{\infty} \phi_t(x_o)\, L(x,x_o)\, dx_o \\ &= \phi_t(x) * L(x)\end{aligned} \tag{3.31}$$

This last expression employs the familiar convolution notation.

IV. Rectangular Geometry

A precise and direct relationship between the converter response and the associated object is the general goal of a radiographic examination. We formulate here an effective and generally useful methodology for rectangular objects with an emphasis on dimensioning.

4.1 KNIFE-EDGE CONVERTER RESPONSE

The neutron absorbing knife-edge object is taken as the starting point in this initial analysis because image blurring is then associated with a readily describable object property. For the conditions discussed in the preceding chapter,the optical density variation associated with the knife-edge object is obtained upon transformation from the following converter response function, Sec. 3.5, normalized for convenience to lie between zero and one:

$$\psi(x) = \frac{1}{2} + \frac{1}{\pi}\tan^{-1}(\lambda x) \tag{4.1}$$

with λ as the neutron radiographic resolution parameter.

Our objective now is to use this function in order to extract selected information about rectangular objects of interest.

4.2 EDGE LOCATION

Consider the important connection between the experimentally fitted edge spread function, Eq. (4.1), and the location of the object's edge. In view of the discussion of Chapter III, if x_+ is the x-coordinate of the knife-edge boundary, Fig. 4.1, then

$$\psi(x) = \frac{1}{2} + \frac{1}{\pi}\tan^{-1}[\lambda(x - x_+)] \tag{4.2}$$

is the relevant edge-spread function. The procedure of fitting Eq. (4.2) to an experimentally obtained optical density variation, using the exposure-density relationship, yields simultaneously the resolution parameter, λ, and the edge location, x_+. However, the edge location can also be determined using more direct methods applicable to edge-spread functions. We discuss three such methods and show that they are exact for a converter response that is precisely of the form of Eq. (4.2).

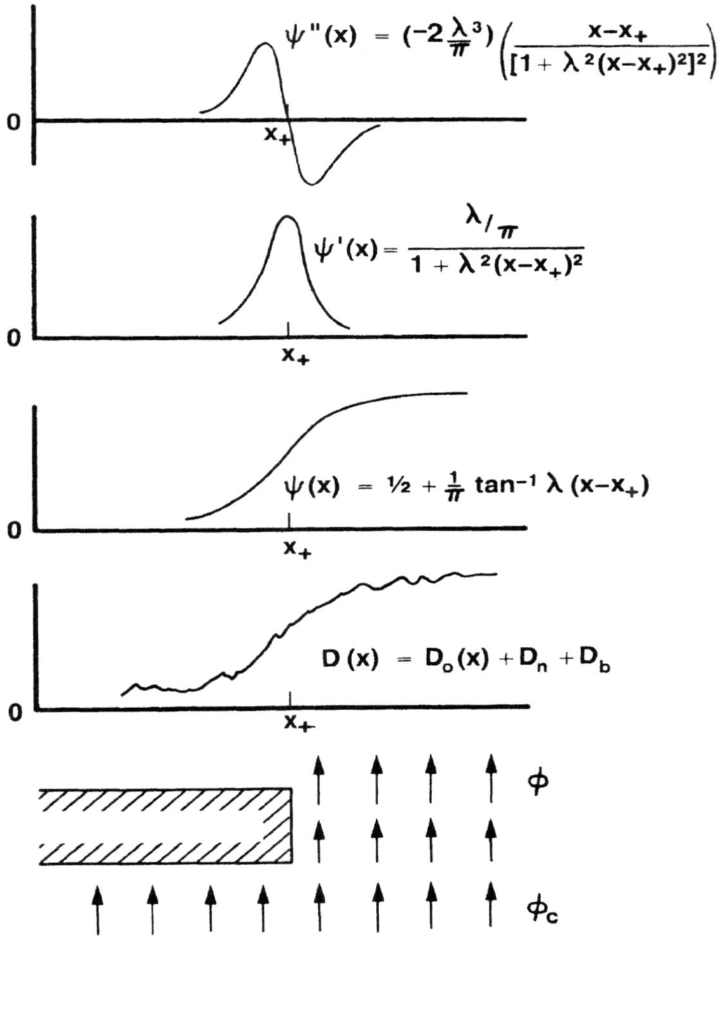

Fig. 4.1: Various related functions associated with an absorbing knife-edge object.

4.2a The $\psi(x_+)$ Criterion

By inspection of Eq. (4.2), with a knowledge of the $\tan^{-1}(\ )$ function, we obtain

$$\psi(x_+) = \lim_{x \to x_+} \{\psi(x)\} = \frac{1}{2} \tag{4.3}$$

That is, for the special case of an ideal neutron absorbing knife-edge object and well collimated neutron beam, the location of the object edge is precisely midway between its asymptotic maximum and asymptotic minimum. In view of the discussion in Sec. 3.6, this criterion is exact when the exposure-density variation is linear.

4.2b The $\psi(x_+)$ Criterion

The first derivative of Eq. (4.2) is

$$\psi'(x) = \frac{\lambda/\pi}{1 + \lambda^2 (x - x_+)^2} \tag{4.4}$$

This function is also the line-spread function centered at $x = x_+$ with a maximum at $x = x_+$ given by

$$\psi'(x_+) = \frac{\lambda}{\pi} \tag{4.5}$$

It follows, therefore, that for a linear exposure-density variation, the slope of the experimentally measured optical density variation attains its maximum value at the precise location of the edge in the image of the ideal knife-edge object. Removal of the noise component prior to differentiation is essential because differentiation of a noisy curve is generally an unstable process.

4.2c The $\psi''(x_+)$ Criterion

Taking a further derivative of Eq. (4.4) yields

$$\psi''(x) = \left(-\frac{2\lambda^3}{\pi}\right) \frac{(x - x_+)}{\left[1 + \lambda^2 (x - x_+)^2\right]^2} \tag{4.6}$$

From this, we find that

$$\psi''(x_+) = 0 \tag{4.7}$$

and therefore conclude that the second derivative of the converter response to the ideal knife-edge is precisely zero at the location of the object's edge.

Figure 4.1 provides a graphical depiction of the three criteria -- $\psi(x)$, $\psi'(x)$ and $\psi''(x)$ -- and how they identify the edge location.

In summary, the edge of an ideal neutron absorbing knife-edge object is located at a point of the converter response where

a) the converter response is midway between the asymptotic values,

b) the converter response gradient is a maximum, or equivalently,

c) the second derivative of the converter response is zero.

We will label b) and c) above as the "Extremum-Slope" Criterion.

These criteria also hold for the optical density directly if the exposure-density variation is in the linear range and indirectly, that is requiring a transformation, Sec. 3.6, otherwise. However, these three criteria must hold simultaneously; if they do not, then the necessary experimental conditions of Sec. 4.1 do not hold or the experimental microdensitometer trace has been improperly obtained with respect to data smoothing, background subtraction, or nonlinearity of the microdensitometer system.

As a detailed, though conceptually important point, we note that the actual numerical value of the resolution parameter did not enter into the definition of the three criteria for the edge location. Hence, the edge of an ideal knife-edge object can be determined without a knowledge of the numerical value of λ, thereby avoiding curve-fitting procedures.

Further thought suggests that the previously stated requirements on exposure-density linearity can be relaxed as follows for an absorbing knife-edge object: the linearity requirement needs only to apply to the optical density in the midway vicinity of the optical density range, being applicable only over an x-coordinate interval sufficient for the determination of the first and second derivatives of the optical density variation. Indeed, over a sufficiently small interval, a log-linear variation is linear to a degree of approximation. These criteria may, however, not necessarily hold for other more complex object geometries.

4.3 WIDE-SLAB OBJECT

A knowledge of how to determine the locations of an edge in the radiographic image of an ideal knife-edge object can be directly used to determine the dimension of a neutron absorbing slab object. If the slab is at least several FWHM wide, the two experimentally obtained converter response functions associated with the left ($x = x_-$) and the right ($x = x_+$) edges are, to a good approximation, independent, Fig. 4.2. Indeed, these two functions are mirror images and are given by

$$\psi_-(x) = \frac{1}{2} - \frac{1}{\pi} \tan^{-1}[\lambda(x - x_-)] \tag{4.8a}$$

and

$$\psi_+(x) = \frac{1}{2} + \frac{1}{\pi} \tan^{-1}[\lambda(x - x_+)] \tag{4.8b}$$

These two response functions are only applicable in the vicinity of the edges they respectively describe. The slab converter response valid for all x and all widths is actually the sum of these two components, given by

$$\psi(x) = \psi_-(x) + \psi_+(x)$$

$$= 1 + \frac{1}{\pi}\{\tan^{-1}[\lambda(x - x_+)] - \tan^{-1}[\lambda(x - x_-)]\} \tag{4.9}$$

The determination of the width of the slab, w, involves the determination of x_+ and x_-. Since the slab is of sufficient width, in terms of FWHM, the right edge coordinate, x_+, can be determined by applying the three converter response criteria derived in the previous section. The left edge coordinate, x_-, is similarly determined by applying the same criteria to $\psi_-(x)$, with the exception that $\psi'_-(x)$ is a minimum at $x = x_-$. The relationship between the criteria for determining x_+ and x_- is depicted in Fig. 4.2. For the wide-slab object, both x_+ and x_- are thus determined independently and the width of the slab is given by $w = x_+ - x_-$.

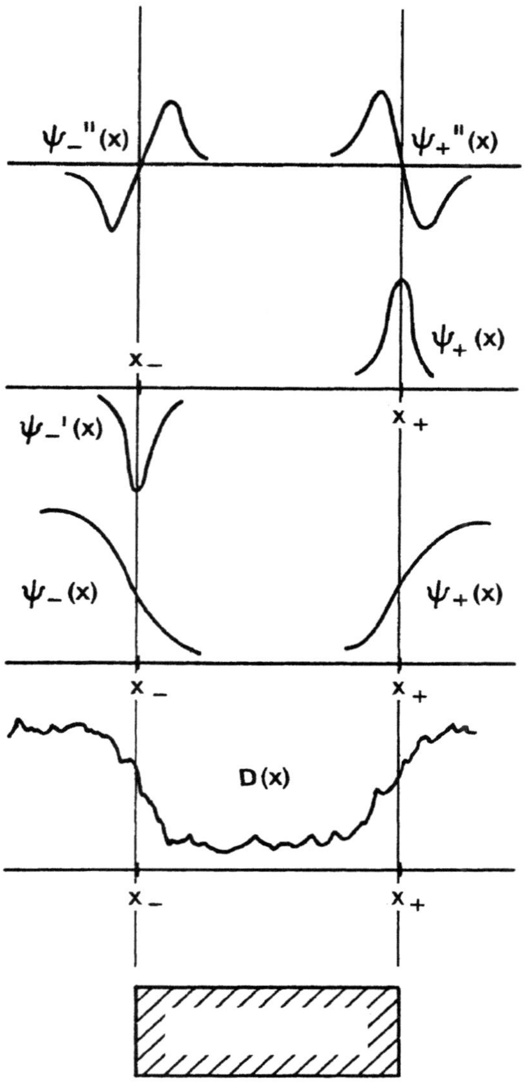

Fig. 4.2: Various related functions associated with an absorbing wide-slab object.

4.4 WIDE-GAP OBJECT

The geometrical inverse of the wide-slab is the wide-gap, as shown in Fig. 4.3. Here, in contrast to Eqs. (4.8), we have

$$\psi_-(x) = \frac{1}{2} + \frac{1}{\pi}\tan^{-1}[\lambda(x - x_-)] \tag{4.10a}$$

and

$$\psi_+(x) = \frac{1}{2} - \frac{1}{\pi}\tan^{-1}[\lambda(x - x_+)] \tag{4.10b}$$

These functions are valid as converter responses in the vicinity of the edges they describe. Note the reversal of (+) and (–) signs for this gap in comparison to the description of the wide-slab, Sec. 4.3. Evidently, as for the wide-slab object, the criteria for the left and right edges are unchanged, except for the sign, and are found to be as follows

$$\psi_-(x_-) = \psi_+(x_+) = \frac{1}{2} \tag{4.11a}$$

$$\psi_-(x_-) \Rightarrow \max; \quad \psi'_+(x_+) \Rightarrow \min \tag{4.11b}$$

$$\psi''_-(x_-) = \psi''_+(x_+) = 0 \tag{4.11c}$$

The associated graphical depictions are also given in Fig. 4.3. No new requirements are introduced which were not already relevant for the wide-slab object. The actual density variation, valid for a gap of arbitrary width, is given by

$$\psi(x) = \psi_+(x) + \psi_-(x) - 1$$

$$= \frac{1}{\pi}\{\tan^{-1}[\lambda(x - x_-)] - \tan^{-1}[\lambda(x - x_+)]\} \tag{4.12}$$

4.5 NARROW-SLAB AND NARROW-GAP OBJECTS

The extraction of dimensional information from the radiographs of neutron absorbing rectangular objects such as the knife-edge, wide-slab, and wide-gap, has been described with the direct use of the edge-spread function, Eq. (4.1). The approach employed for these objects has the attractive property that edge locations can be determined without reference to either

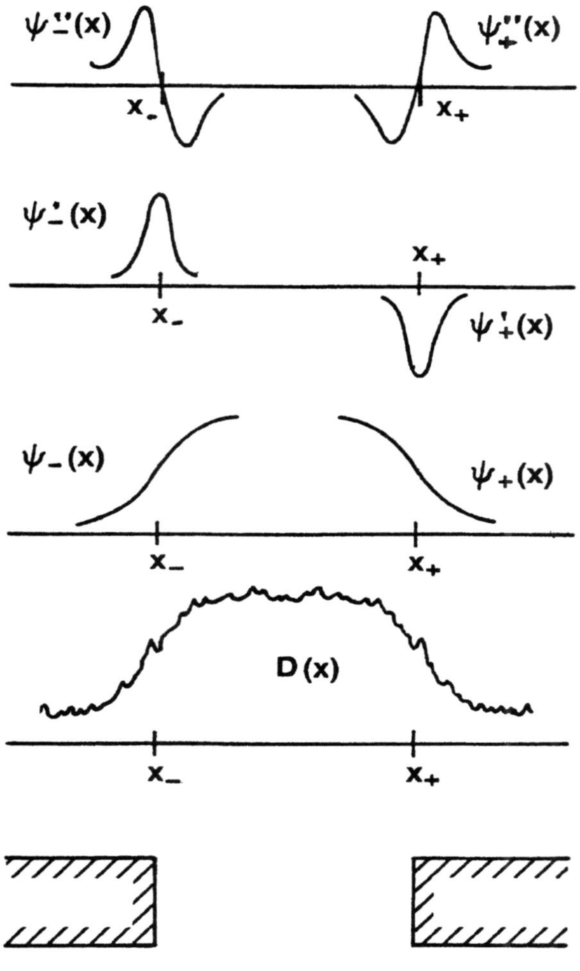

Fig. 4.3: Various related functions associated with a wide-gap in an absorbing object.

the system line-spread function or the object itself. This approach is effective in providing an intuitive understanding to radiographic unsharpness analysis.

It is, however, only for the above elementary objects with isolated or widely separated edges that such a simplified approach is possible. In general, the convolution approach suggested in Sec. 3.6 needs to be employed and edge location prescriptions may be complex depending on the object itself and the system line-spread function. We consider this approach and apply it to the three elementary objects of the preceding sections with the previous restriction of widely separated edges removed.

As derived in Sec. 3.6, the general converter response along the x-direction is fully described by the convolution integral

$$\psi(x) = N \int_{-\infty}^{\infty} \phi_t(x_0)\, L(x, x_0)\, dx_0 \tag{4.13}$$

with N as a convenient normalization constant. Here, $\phi_t(x_0)$ is the intensity -- either flux or fluence -- of the neutron beam which has penetrated the object at x_0. Further, a suitable line-spread function, Sec. 3.3, is

$$L(x,x_0) = \frac{\lambda}{\pi}\,[1 + \lambda^2(x - x_0)^2]^{-1} \tag{4.14}$$

We now apply the general response in Eq. (4.13) to the three specific elementary objects of the preceding section with object width restrictions removed.

4.5a The Totally Absorbing Knife-Edge Object

With the edge located at $x = x_+$, Fig. 4.1, it is convenient to represent the convolution integral, Eq. (4.13), over two distinct domains as follows:

$$\psi(x) = N \int_{-\infty}^{x_+} \phi_t(x_0)\, L(x, x_0)\, dx_0 + N \int_{x_+}^{\infty} \phi_t(x_0)\, L(x, x_0)\, dx_0 \tag{4.15}$$

This equation is similar to the convolution equation for the edge-spread function in Sec. 3.6.In the first interval, $(-\infty, x_+)$, neutron absorption is total and hence none of the incident neutrons are transmitted. For the second interval, (x_+, ∞), there exists no neutron absorption and all the incident neutrons impinge on the converter. Hence, for this case we have

$$\phi_t(x_o) = \begin{cases} 0, & x_o \le x_+ \\ \phi_o, & x_o > x_+ \end{cases} \tag{4.16}$$

and Eq. (4.15) reduces to

$$\psi(x) = N\,\phi_o \int_{x_+}^{\infty} \left\{ \frac{\lambda/\pi}{1+\lambda^2(x-x_o)^2} \right\} dx_o$$

$$\propto \frac{1}{2} + \frac{1}{\pi} \tan^{-1}[\lambda(x-x_+)] \tag{4.17}$$

This last expression for the converter response associated with the neutron absorbing knife-edge object, Figs. 4.1, is precisely that given in Eq. (4.1).

4.5b The Totally Absorbing Narrow-Slab Object

For the slab of Fig. 4.2, the incident beam is described by

$$\phi_t(x_o) = \begin{cases} \phi_o, & x_o < x_- \\ 0, & x_- \le x_o \le x_+ \\ \phi_o, & x_o > x_+ \end{cases} \tag{4.18}$$

and we write for the three distinct domains

$$\psi(x) = N \int_{-\infty}^{x_-} \phi_t(x_o)\, L(x, x_o)\, dx_o + N \int_{x_-}^{x_+} \phi_t(x_o)\, L(x, x_o)\, dx_o + N \int_{x_+}^{\infty} \phi_t(x_o)\, L(x, x_o)\, dx_o$$

$$= N\phi_0 \int_{-\infty}^{x_-} \left\{ \frac{\lambda/\pi}{1+\lambda^2(x-x_o)^2} \right\} dx_o + 0 + N\,\phi_0 \int_{x_+}^{\infty} \left\{ \frac{\lambda/\pi}{1+\lambda^2(x-x_o)^2} \right\} dx_o \tag{4.19}$$

Integration yields

$$\psi(x) \propto 1 + \frac{1}{\pi} \{\tan^{-1}[\lambda(x-x_+)] - \tan^{-1}[\lambda(x-x_-)]\} \tag{4.20}$$

This is precisely the expression derived in Sec. 4.3 and presented in Eq. (4.9). If the slab width is several FWHM, the two components of Eq. (4.19) do not overlap and, like Eqs. (4.8a) and

and (4.8b) may be used independently to represent $\psi(x)$ near their respective edges. If the slab is narrow, then the two components overlap and the entire equation, that is both components of Eq. (4.20), must be used in any experimental dimensional measurement.

This latter case creates some complications for dimensioning purposes since there exists now one relation, Eq. (4.20), but two unknowns, x_- and x_+. The distinction between wide-slab and narrow-slab converter responses is displayed in Fig. 4.4. A dimensioning procedure can, however, be devised for the narrow-slab object that requires only the maximum and minimum converter responses without knowledge of the resolution parameter, λ. From the proportionality expression, Eq. (4.20), the maximum and minimum converter responses appear in the following ratio:

$$\frac{\Psi_{min}}{\Psi_{max}} = \psi\left(\frac{x_+ + x_-}{2}\right) / \psi(\infty)$$

$$= 1 - \frac{2}{\pi}\tan^{-1}\left[\lambda\left(\frac{x_+ - x_-}{2}\right)\right] \tag{4.21}$$

Note that this equation can be inverted to

$$\lambda(x_+ - x_-) = 2\tan\left[\frac{\pi}{2}\left(1 - \frac{\Psi_{min}}{\Psi_{max}}\right)\right] \tag{4.22}$$

Likewise, the edge converter responses are given by the ratios

$$\frac{\psi(x_+)}{\Psi_{max}} = \frac{\psi(x_-)}{\Psi_{max}} = 1 - \frac{1}{\pi}\tan^{-1}[\lambda(x_+ - x_-)] \tag{4.23}$$

By combining Eqs. (4.22) and (4.23), one finds that the edges can be identified as those image locations given by

$$\psi(x_+) = \psi(x_-) = \psi_{max}\left[1 - \frac{1}{\pi}\tan^{-1}\left\{2\tan\left[\frac{\pi}{2}\left(1 - \frac{\Psi_{min}}{\Psi_{max}}\right)\right]\right\}\right] \tag{4.24}$$

While possessing functional dependence only on ψ_{min} and ψ_{max}, this expression is considerably more complex than the midway converter response prescription valid for a wide-slab object. Note that in the limit of a wide-slab object, ψ_{min} approaches zero and Eq. (4.24) reduces to the simpler wide-slab prescription.

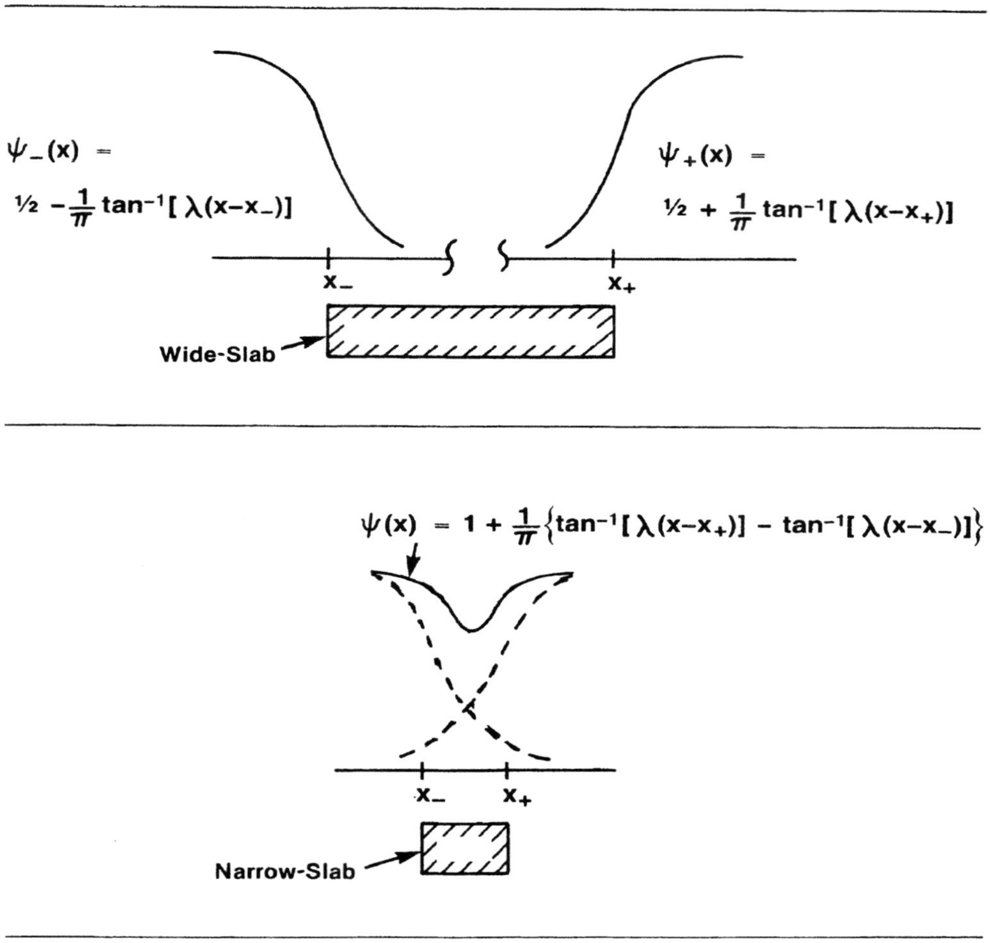

Fig. 4.4: Illustration showing the different converter responses for wide-slab and narrow-slab absorbing objects.

4.5c The Totally Absorbing Narrow-Gap Object

In Sec. 4.4 we had considered the wide-gap object, Fig. 4.3, so that the edge-spread functions associated with each edge could be considered independently. We next choose not to place any dimensional restrictions on the gap and use, for the neutron beam impinging on the converter, the intensity

$$\phi_t(x_o) = \begin{cases} 0, & x_o < x_- \\ \phi_o, & x_- \leq x_o \leq x_+ \\ 0, & x_o > x_+ \end{cases} \tag{4.25}$$

Then according to the general expression, Eq. (4.13), and $\phi_t(x_o)$ as specified, we obtain

$$\psi(x) = N \int_{-\infty}^{\infty} \phi_t(x_o)\, L(x, x_o)\, dx_o$$

$$= N \phi_o \int_{x_-}^{x_+} \left\{ \frac{\lambda/\pi}{1 + \lambda^2 (x - x_o)^2} \right\} dx_o$$

$$\propto \frac{1}{\pi} \{\tan^{-1}[\lambda(x - x_-)] - \tan^{-1}[\lambda(x - x_+)]\} \tag{4.26}$$

This is precisely the expression derived in Sec. 4.4 and presented in Eq. (4.12). The distinction between the narrow-gap and wide-gap converter responses is displayed in Fig. 4.5. Unlike the narrow-slab object, however, it is not possible to derive an expression for the narrow-gap edge location converter response as a function of easily identifiable converter responses or their derivatives. This arises because the only readily identifiable non-zero converter response is the maximum, ψ_{max}, whereas both ψ_{max} and ψ_{min} are available in the narrow-slab case.

4.6 PARTIALLY ABSORBING OBJECTS

Direct methods for locating the edges in radiographic images of knife-edge, slab and gap objects were derived in previous sections under suitable assumptions. Notable among these was the constraint that the rectangular objects be "neutronically black", absorbing all neutrons impingent on them. This restriction was, however, introduced to prevent the mathematical description from unnecessarily obscuring the essential imaging physics. Indeed, the restrictive assumption of a totally absorbing object can be relaxed to allow objects of a constant, finite absorption cross section, Σ_a, without changing any of the previous edge location criteria.

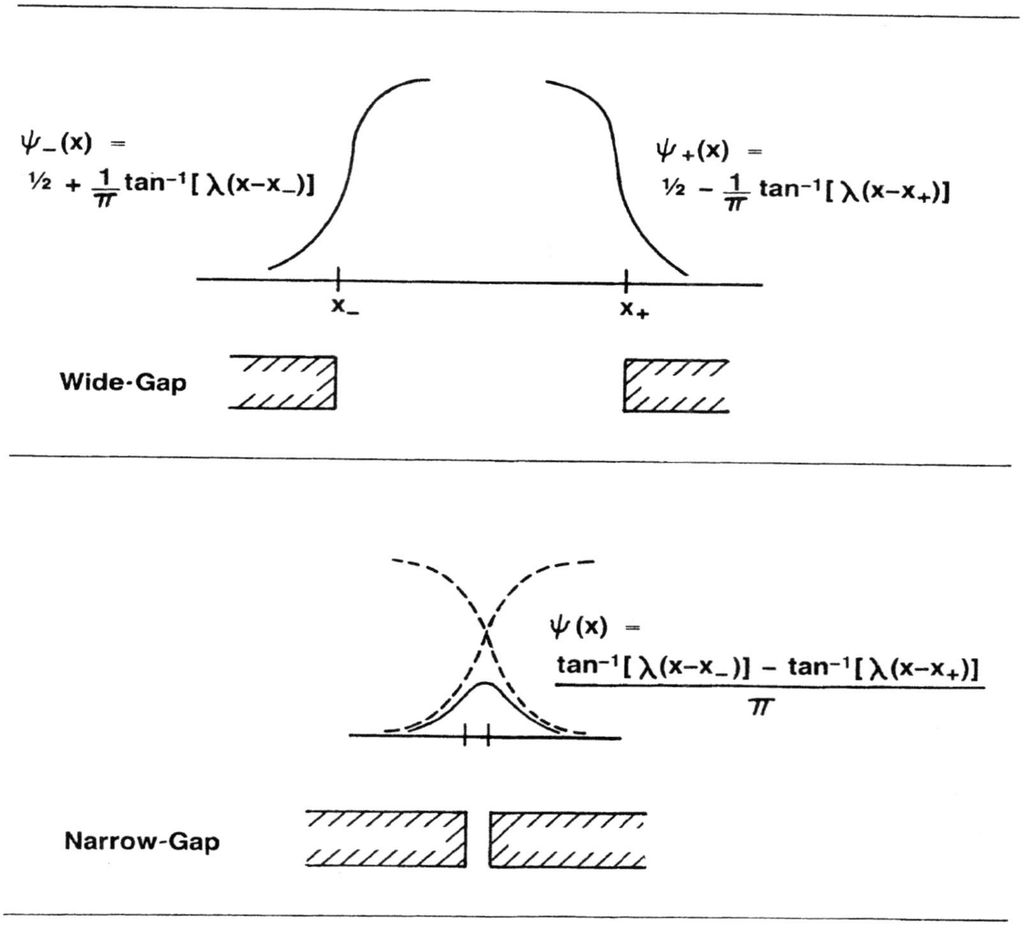

Fig. 4.5: Illustration showing the different converter responses for a wide-gap and narrow-gap in an absorbing object.

If the knife-edge slab object of Fig. 4.1 is homogeneous with Σ_a as the absorption cross section and thickness Δz, then the neutron beam exiting the slab object has an intensity, Sec. 2.3, given by

$$\phi_t(x_o) = \begin{cases} \phi_o \exp[-\Sigma_a \Delta z],\ x_o \le x_+ \\ \phi_o,\ x_o > x_+ \end{cases} \tag{4.27}$$

In this case the convolution integral, Eq. (4.13), becomes

$$\psi(x) \propto \int_{-\infty}^{x_+} \exp[-\Sigma_a \Delta z] \left\{ \frac{\lambda/\pi}{1 + \lambda^2 (x - x_o)^2} \right\} dx_o + \int_{x_+}^{\infty} \left\{ \frac{\lambda/\pi}{1 + \lambda^2 (x - x_o)^2} \right\} dx_o \tag{4.28}$$

Performing the above integration yields the edge-spread function

$$\psi(x) \propto (1 + \exp[-\Sigma_a \Delta z]) \frac{1}{2} + \left(\frac{1 - \exp[-\Sigma_a \Delta z]}{\pi} \right) \tan^{-1}[\lambda(x - x_+)] \tag{4.29}$$

which is similar in form and reduces to the corresponding totally absorbing knife-edge expression, Eq. (4.17), when $\Sigma_a \rightarrow \infty$. Now the maximum, minimum and edge coordinate converter responses are

$$\psi(\infty) = 1 \text{ (maximum at } x = \infty) \tag{4.30a}$$

$$\psi(x_+) = (1 + \exp[-\Sigma_a \Delta z])/2 \text{ (at edge)} \tag{4.30b}$$

$$\psi(-\infty) = \exp[-\Sigma_a \Delta z] \text{ (minimum at } x = -\infty) \tag{4.30c}$$

with $\psi(x_+)$ clearly lying halfway between the maximum and minimum responses. In similarity to Sec. 4.2, it also follows that $\psi'(x)$ is a maximum and $\psi''(x)$ vanishes at $x = x_+$. The three edge location criteria of Sec. 4.2 thus hold for a partially absorbing knife-edge object as well. Further, it is a simple extension to show that these criteria can be applied in the dimensioning of partially absorbing slab and gap objects, in the manner described in Secs. 4.3 and 4.4.

V. Cylindrical Geometry

The mathematical analysis of the converter response for radiographed cylindrical objects introduces considerations which were not encountered with rectangular specimens. We assess these effects and again emphasize the extraction of dimensional information.

5.1 HOMOGENEOUS CYLINDER

The experimental optical density variation associated with a cylindrical object together with the incident and transmitted neutron beam is shown in Fig. 5.1. Based on our previous discussion, Sec. 3.6 the converter response variation, $\psi(x)$, associated specifically with this cylindrical object is given by the following general convolution expression:

$$\psi(x) = N \int_{-\infty}^{\infty} \phi_t(x_0)\, L(x,x_0)\, dx_0 \tag{5.1}$$

Here, $\phi_t(x_0)$ is the transmitted neutron beam intensity which, for the homogeneous cylinder with absorption as the dominant neutron-nucleus interaction, is given by

$$\phi_t(x_0) = \begin{cases} \phi_0, & x < x_- \\ \phi_0 \exp[-\Sigma_a Z(x_0)], & x_- \le x \le x_+ \\ \phi_0, & x > x_+ \end{cases} \tag{5.2}$$

Note that here the macroscopic absorption cross section, Σ_a, is constant since the cylinder is taken to be homogeneous, and $Z(x_0)$ is the neutron path length through the object, as depicted in Fig. 5.1. Implicit in this one-dimensional formulation is the assumption that the cylinder is of sufficient axial length that there is negligible axial variation in the resulting converter response.

Substitution of Eq. (5.2) into Eq (5.1) leads to three distinct regions of integration over the x_0 variable:

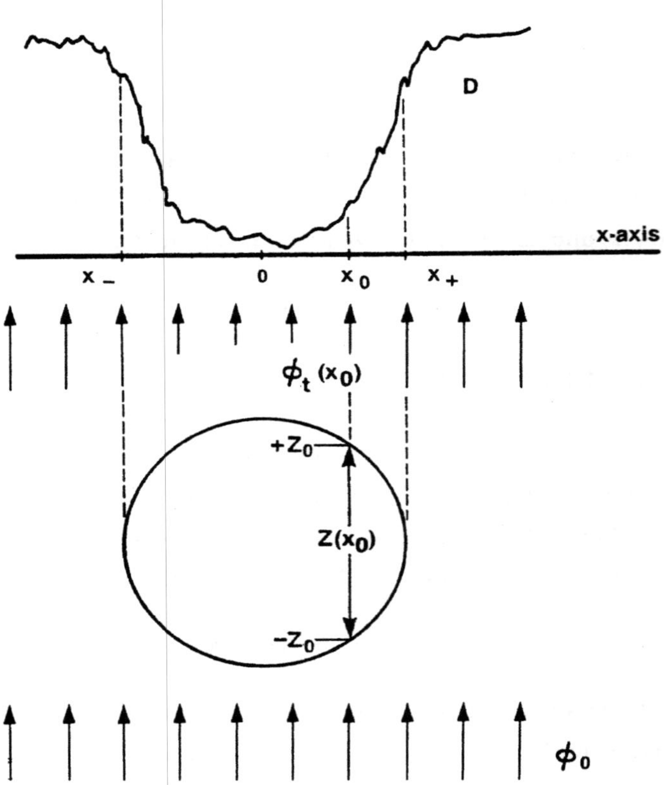

Fig. 5.1: Depiction of incident and transmitted neutron beam and associated optical density variation for a cylindrical object.

$$\psi(x) = N\phi_o \int_{-\infty}^{x-} L(x, x_o)\, dx_o$$
$$+ N\phi_o \int_{x-}^{x+} \exp[-\Sigma_a Z(x_o)]\, L(x, x_o)\, dx_o \qquad (5.3)$$
$$+ N\phi_o \int_{x+}^{\infty} L(x, x_o)\, dx_o$$

This expression is similar to its counterpart for the partially absorbing slab, Sec. 4.6, with the important difference that the attenuation function in the central integrand is not constant. Application of the line-spread function appropriate for our purposes

$$L(x, x_o) = \frac{\lambda/\pi}{1 + \lambda^2 (x - x_o)^2} \qquad (5.4)$$

yields specifically

$$\psi(x) = N\phi_o \left\{ \frac{1}{2} - \frac{1}{\pi} \tan^{-1}[\lambda (x - x_-)] \right\}$$
$$+ N\phi_o \frac{\lambda}{\pi} \int_{x_-}^{x_+} \frac{\exp[-\Sigma_a Z(x_o)]}{1 + \lambda^2 (x - x_o)^2}\, dx_o \qquad (5.5)$$
$$+ N\phi_o \left\{ \frac{1}{2} + \frac{1}{\pi} \tan^{-1}[\lambda (x - x_+)] \right\}$$

A physical interpretation of Eq. (5.5) is that the converter response, and hence the optical density variation, from a cylinder consists of two paired components associated with a perfectly absorbing ideal knife-edge and a third component associated with neutron attenuation in cylindrical geometry. The separate knife-edge components and the central cylindrical component can be illustrated by taking the origin of the x-axis at the midway point between x_- and x_+ and specializing Eq. (5.5) to the half-space $x \in [0, \infty)$ to give

$$\psi_+(x) = \frac{1}{2} + \frac{1}{\pi} \tan^{-1}[\lambda (x - x_+)]$$
$$+ \frac{\lambda}{\pi} \int_{0}^{x_+} \frac{\exp[-\Sigma_a Z(x_o)]}{1 + \lambda^2 (x - x_o)^2}\, dx_o \qquad (5.6)$$

Here the normalization of $N = 1/\phi_o$ has been selected and the left-hand edge-spread term, as well as the lower half of the integral, have been omitted. These omissions do not detract from the accuracy if the cylinder diameter is at least several FWHM of the line-spread function.

We display, in Fig. 5.2, each of these components based on a numerical evaluation of this equation using parameters of typical interest in the neutron radiographic examination of nuclear fuel pins, albeit with Σ_a as a constant. Any considerations for determining the precise location of the cylindrical edge must include the integral term of Eq. (5.6) because it contributes considerably to the localized edge effect.

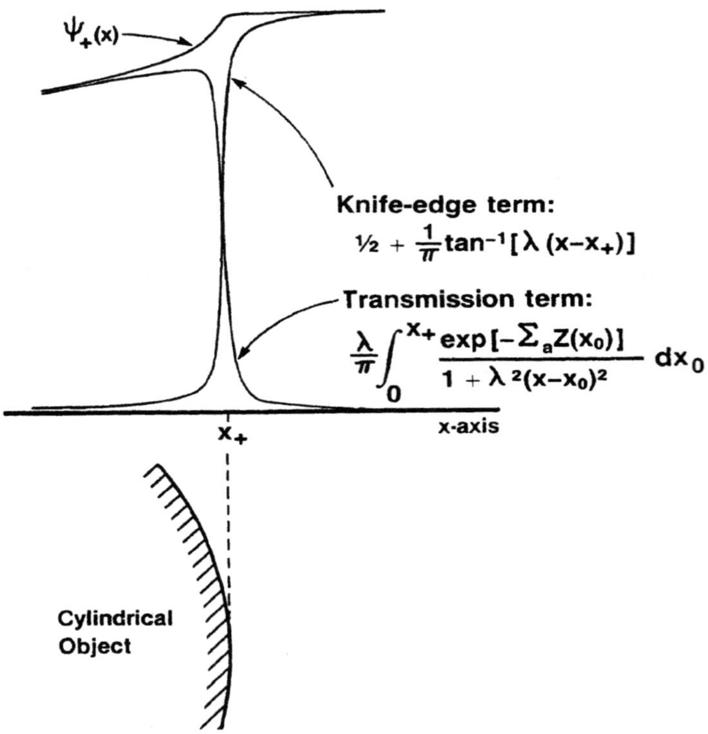

Fig. 5.2: Converter response, $\psi_+(x)$, and its two components for a homogeneous cylindrical object.

5.2 CYLINDER-EDGE LOCATION

The diameter of a cylinder could be obtained if the two edge coordinates, x_- and x_+, could be determined from the optical density trace. We therefore investigate the identification of properties of $\psi(x)$ about an edge, which could then be used to locate x_- or x_+

in a manner similar to that used for the knife-edge object of the preceding chapter; that is, we consider the extremum-slope criterion of Ch. IV, now applied in cylindrical geometry.

First, we consider specifically the x_+ edge, Fig. 5.1 and Eq. (5.6). Then, supposing the cylindrical object is of a radius R (i.e. $x_+ = R$), the optical path of the transmitted neutron beam at an arbitrary coordinate x_o $(0 < x_o < x_+)$ is evidently

$$Z(x_o) = 2\sqrt{R^2 - x_o^2} \tag{5.7}$$

The converter response for the $x \geq 0$ domain is therefore given by Eq. (5.6) with $x_+ = R$ and the substitution of Eq. (5.7):

$$\psi_+(x) = \frac{1}{2} + \frac{1}{\pi}\tan^{-1}[\lambda(x - R)] + \frac{\lambda}{\pi}\int_o^R \frac{\exp[-2\Sigma_a(R^2 - x_o^2)^{1/2}]}{1 + \lambda^2(x - x_o)^2}\, dx_o \tag{5.8}$$

The optical density at the edge $x = R$ is then

$$\psi_+(R) = \frac{1}{2} + \frac{\lambda}{\pi}\int_o^R \frac{\exp[-2\Sigma_a(R^2 - x_o^2)^{1/2}]}{1 + \lambda^2(R - x_o)^2}\, dx_o$$

$$= \frac{1}{2} + fc(R, \Sigma_a, \lambda), \tag{5.9}$$

implicitly defining the positive function, fc. Thus, in contrast to the ideal knife-edge case, the converter response at the cylindrical edge exceeds 1/2 of the minimum-to-maximum converter response and depends upon the radius, R, the macroscopic absorption cross section, Σ_a, as well as the resolution parameter, λ. Hence, simple edge location criteria, like those derived in Sec. 4.2 for rectangular objects, are not possible for a cylinder.

We next consider the derivative of $\psi_+(x)$ about the edge at $x = R$. Based on Leibnitz's rule of differentiation under an integral, it can be shown that

$$\frac{d\psi_+(x)}{dx} = \frac{\lambda/\pi}{1 + \lambda^2(x - R)^2} - \frac{2\lambda^3}{\pi}\int_o^R \frac{(x - x_o)\exp[-2\Sigma_a(R^2 - x_o^2)^{1/2}]}{[1 + \lambda^2(x - x_o)^2]^2}\, dx_o \tag{5.10}$$

Evaluating this expression at $x = R$ gives

$$\psi_+(R) = \frac{\lambda}{\pi}\left\{1 - 2\lambda^2 \int_0^R \frac{(R-x_0)\exp[-2\Sigma_a(R^2-x_0^2)^{1/2}]}{[1+\lambda^2(R-x_0)^2]^2}\,dx_0\right\} \tag{5.11}$$

This expression cannot be appreciably simplified analytically so that the first derivative is also not useful in providing a compact edge location method.

The second derivative of Eq. (5.8) follows by differentiation of Eq. (5.11) once more and yields

$$\frac{d^2\psi_+(x)}{dx^2} = \frac{2\lambda^3}{\pi}\left\{\frac{(R-x)}{[1+\lambda^2(x-R)^2]^2} - \int_0^R \frac{\exp[-2\Sigma_a(R^2-x_0^2)^{1/2}]}{[1+\lambda^2(x-x_0)^2]^3}[1-3\lambda^2(x-x_0)^2]\,dx_0\right\} \tag{5.12}$$

At $x = R$, this simplifies somewhat to

$$\psi_+(R) = \frac{-2\lambda^3}{\pi}\int_0^R \frac{\exp[-2\Sigma_a(R^2-x_0^2)^{1/2}]}{[1+\lambda^2(R-x_0)^2]^3}[1-3\lambda^2(R-x_0)^2]dx_0 \tag{5.13}$$

A detailed examination of these functional relationships reveals that mathematical operations on $\psi_+(x)$ provide apparently little help in identifying a simple and convenient methodology for the determination of the edge in a cylinder image. A numerical evaluation, however, reveals a consistent bias between the actual radius of a cylinder, R, and that obtained using the extremum slope criterion. The extremum-slope estimate of the radius, R_{ext}, is a solution of

$$\left.\frac{d^2\psi_+(x)}{dx^2}\right|_{x=R_{ext}} = 0 \tag{5.14}$$

A sample calculational result is displayed in Table 5.1 for a range of R of common interest and for $\Sigma_a = 0.2\ cm^{-1}$ and $\lambda = 25mm^{-1}$. These numerical values generally correspond to those of interest in the neutron radiographic examination of fresh nuclear reactor fuel pins.

A study of Table 5.1 suggests two important points: one is that the extremum-slope edge underestimates the actual location of the cylindrical edge, depending upon the effect of the exposure-density of the film, and the other is that this underestimate is only weakly dependent on the radius. Indeed, the precision here displayed suggests that an excellent estimate for the cylindrical edges be obtained by adding the small correction ΔR

$$R_{est} = R_{ext} + \Delta R \tag{5.15}$$

We have further explored the range of ΔR as a function of R, Σ_a and λ and display these results in Table 5.2 Note here that ΔR again possesses only a weak dependence on Σ_a and that it is primarily a function of the resolution parameter, λ.

Table 5.1

Calculational results of R_{ext} and comparison to actual radius, R, for $\Sigma_a = 0.2\ cm^{-1}$ and $\lambda = 25\ mm^{-1}$

R_{ext} (mm)	R (mm)	$\Delta R = (R - R_{ext})$ (mm)
2.9776	3.00000	0.0224
3.9776	4.00000	0.0224
4.9776	5.00000	0.0224
5.9776	6.00000	0.0224
6.9777	7.00000	0.0223
7.9777	8.00000	0.0223
8.9777	9.00000	0.0223

5.3 INHOMOGENEOUS CYLINDER

Inhomogeneous cylinders are characterized by a non-uniform isotope distribution, therefore resulting in a macrosopic cross section which is variable throughout the object. The most common spatial variations exhibit circular symmetry, for which

$$\Sigma_a \rightarrow \Sigma_a(r) \tag{5.16}$$

where r denotes the radial distance from the axis of the cylinder. Evidently, a hollow cylinder is simply characterized by $\Sigma_a = 0$ with a specified radial domain.

Table 5.2

Tabulation of average ΔR ($= R - R_{ext}$) for $3 \leq R \leq 9$ mm; ΔR are in μm.

		λ (mm^{-1})		
		10	25	50
Σ_a (cm^{-1})	0.2	54.5 ± 0.1	22.3 ± 0.1	11.3 ± 0.1
	0.4	54.6 ± 0.5	21.8 ± 0.2	11.1 ± 0.1

The analysis of the preceding sections now needs to be extended to incorporate radial variations in the cross section. An examination of Eq. (5.3) reveals that the absorption cross section appears only in the exponential attenuation along the $Z(x_0)$ traverse of the cylinder, as indicated in Fig. 5.1. Hence, the attenuation term in Eq. (5.3) must be replaced by the equivalent integration along the traverse as follows:

$$\Sigma_a Z(x_0) \rightarrow \int_{-Z_0}^{Z_0} \Sigma_a(r)\, dz \tag{5.17}$$

By extension of Eq. (5.3) we write the following expression for the computed converter response associated with an inhomogeneous cylinder:

$$\begin{aligned} \psi(x) &= N\phi_0 \int_{-\infty}^{x_-} L(x, x_0)\, dx_0 \\ &+ N\phi_0 \int_{x_-}^{x_+} \exp\left[- \int_{-Z_0}^{Z_0} \Sigma_a(r)\, dz \right] L(x, x_0)\, dx_0 \\ &+ N\phi_0 \int_{x_+}^{\infty} L(x, x_0)\, dx_0 \end{aligned} \tag{5.18}$$

Though a number of geometrical relationships can be invoked to rewrite the inner integral term of this equation, analytical evaluations even for simplified Σ_a (r) expressions do not exist. However, some special cases can be defined and evaluated numerically.

The case of a hollow cylinder is suggested in Fig. 5.3. Heterogeneity consists here of two radial regions for which the following holds:

$$\begin{aligned} \Sigma_a(r) &= 0 , & r &< R_1 \\ \Sigma_a(r) &= \Sigma_a , & R_1 &< r < R_2 \end{aligned} \tag{5.19}$$

The evaluation of the central term in Eq. (5.18) is thereby simplified.

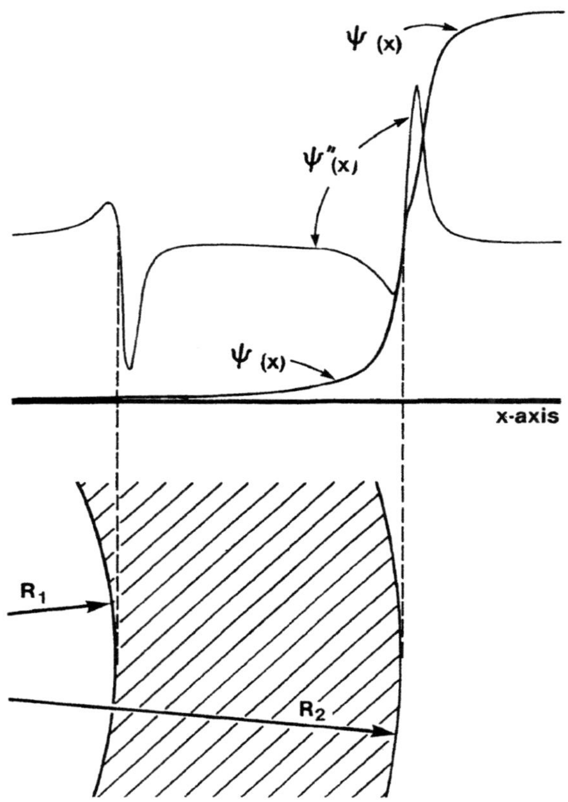

Fig. 5.3: Converter response, $\psi(x)$, and its second derivative, $\psi''(x)$, for a hollow cylinder $R_1 = 5.0$ mm, $R_2 = 7.5$ mm, $\Sigma_a = 0.02\ cm^{-1}$, $\lambda = 10\ mm^{-1}$.

In Fig. 5.3 we display both the right boundary converter response, $\psi(x)$, and its second derivative, $\psi''(x)$, for a hollow cylinder characterized by an interior radius of $R_1 = 5$ mm and exterior radius of $R_2 = 7.5$ mm. The resolution parameter is $\lambda = 10\ mm^{-1}$ and the material cross section is $\Sigma_a = 0.02\ cm^{-1}$. The significant feature to note here is that $\psi(x)$ displays an additional distortion on the outer edge domain and a barely perceptible effect in the inner edge region. However, $\psi''(x)$ displays very strong features, suggesting again that the second derivative carries the dominant edge location information.

From the zero's of $\psi''(x)$ the extremum-slope radii are found to be

$$R_{1,ext} = 4.9427 \text{ mm} \quad (\therefore \Delta R_1 = 57.3\ \mu m)$$

$$R_{2,ext} = 7.4433 \text{ mm} \quad (\therefore \Delta R_2 = 56.7\ \mu m)$$

which are again underestimated.

The other important form of inhomogeneity involves spatially varying cross sections such as those associated with nuclear fuel pins exposed to a neutron flux in a nuclear reactor. We consider here a single quadratic variation given by

$$\Sigma_a(r) = \Sigma_o + (\Sigma_R - \Sigma_o)\left(\frac{r}{R}\right)^2 \tag{5.20}$$

where Σ_o and Σ_R are the macroscopic absorption cross sections in the center ($r = 0$) and edge ($r = R$), respectively. A typical result showing $\psi(x)$ and $\psi''(x)$ is displayed in Fig. 5.4 for the parameters indicated. Again the extremum-slope criterion underestimates the radius; the difference is found to be $\Delta R = 53.8\ \mu m$.

The above and other numerical evaluations suggest that the extremum-slope criterion consistently underestimates the actual cylindrical radii by amounts that are largely insensitive to actual radius and cross sections, but depend primarily upon the resolution parameter. As a "rule-of-thumb" it appears that corrections suggested in Table 5.2 are generally accurate to within 1 μm. Nonlinearity of the exposure-density variation may conceivably have a significant additional effect unless incorporated in the analysis.

5.4 CYLINDRICAL NUCLEAR FUEL

Nuclear reactor fuel generally consists of cylindrical fuel pellets encased in a metallic cladding. A variable gap between the central fuel pellet and outer cladding is unavoidable though a pliable material may be interposed in order to enhance heat conduction and provide for fuel pellet distortion with reduced pressure effects on the cladding. For neutron radiographic purposes, we may therefore represent nuclear fuel as a 3-media cylinder with the core and each annular region characterized by a particular macroscopic cross section, as suggested in Fig. 5.5.

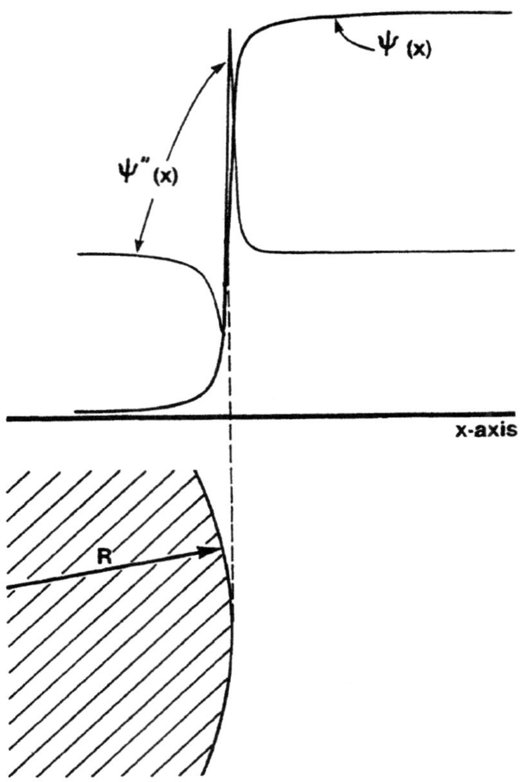

Fig. 5.4: Converter response, $\psi(x)$, and its second derivative, $\psi''(x)$, for a cylindrical object with quadratically varying cross section. $R = 5.0$ mm, $\Sigma_0 = 0.145\ \text{cm}^{-1}$, $\Sigma_R = 0.175\ \text{cm}^{-1}$, $\lambda = 10\ \text{mm}^{-1}$.

In Fig. 5.5 we also show the computed converter response resulting from the evaluation of the convolution integral

$$\psi(x) = N \int_{-\infty}^{+\infty} \phi_t(x_o)\, L(x, x_o)\, dx_o \tag{5.21}$$

Cylindrical symmetry suggests that we write for the right-hand part

$$\psi_+(x) = N\phi_o \int_0^{\infty} \exp\left\{-2\int_0^{Z_o(x_o)} \Sigma_a(z)\, dz\right\} L(x, x_o)\, dx_o \tag{5.22}$$

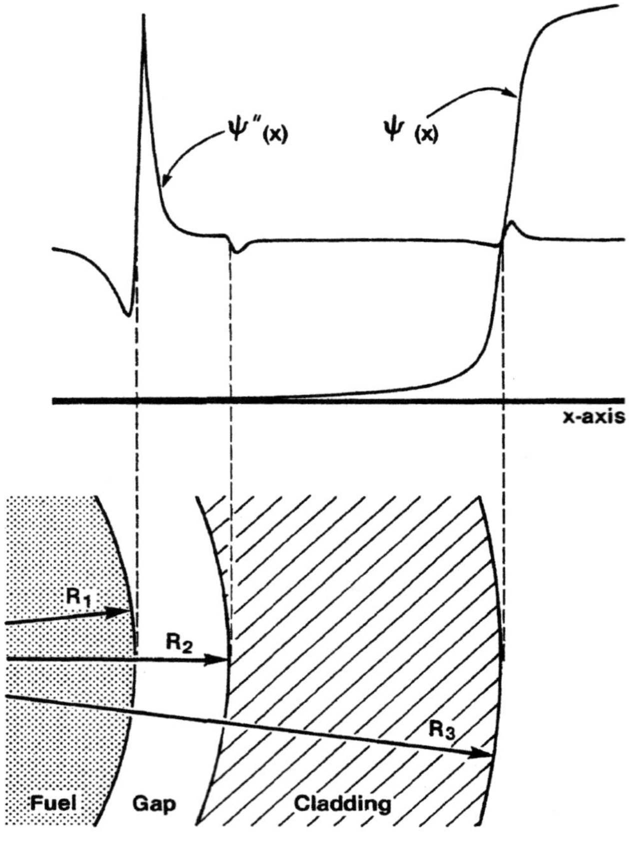

Fig. 5.5: Converter response, $\psi(x)$ and $\psi''(x)$ for a three-media heterogeneous cylinder:
fuel region: $R_1 = 5.0$ mm, $\Sigma_o = 0.145$ cm^{-1}, $\Sigma_R = 0.175$ cm^{-1},
gap region: $R_2 = 6.0$ mm, $\Sigma_a = 0.0001$ cm^{-1},
clad region: $R_1 = 9.0$ mm, $\Sigma_a = 0.01$ cm^{-1},
$\lambda = 100.0$ cm^{-1}.

Here, the upper limit of integration for the inner integral is

$$Z_o(x_o) = Z_1 + Z_2 + Z_3 \tag{5.23}$$

with Z_1, Z_2 and Z_3 representing one half the penetration paths of the incident neutrons at x_o in the 3-media cylinder, Fig. 5.5 . These three path lengths are related to the three radii, R_1, R_2, R_3, at a given x_o by

$$R_1^2 = x_o^2 + Z_1^2 , \qquad x_o \in [0,R_1] \tag{5.24a}$$

$$R_2^2 = x_o^2 + (Z_1 + Z_2)^2 , \qquad x_o \in [0,R_2] \tag{5.24b}$$

$$R_3^2 = x_o^2 + (Z_1 + Z_2 + Z_3)^2 , \quad x_o \in [0,R_3] \tag{5.24c}$$

Thus, for R_1, R_2 and R_3 known, or adequately estimated, the inner integration of Eq. (5.22) can be performed subject to the specification of the macroscopic cross section, $\Sigma_a(z)$, in each of the regions.

A "fresh" nuclear fuel will possess a uniform cross section so that, for such an object,

$$\Sigma_a(z) = \begin{cases} \Sigma_1, & 0 \le Z \le Z_1 \\ \Sigma_2, & Z_1 \le Z \le Z_2 \\ \Sigma_3, & Z_2 \le Z \le Z_3 \end{cases} \tag{5.25}$$

In contrast, a fuel pin exposed to an isotropic neutron flux and radiation fields will acquire a spatially variable cross section with a dependence on the radial distance r. While the cladding and gap-region will likely develop a very weak spatial cross section variation, the fuel region will develop a cross section dependence similar to that given in Eq. (5.20). Accordingly, we make the approximation that a "spent" fuel pin will again have Σ_2 and Σ_3 defined as constants, and that for the inner region Σ_1 can be represented by

$$\Sigma_1 \sim \Sigma_o + [\Sigma_R - \Sigma_o]\left(\frac{r}{R_1}\right)^2 , \quad 0 \le r \le R_1 \tag{5.26}$$

where the radial coordinate r now identifies the integration variable z by

$$r^2 = z^2 - x_o^2 \tag{5.27}$$

Thus, both a "fresh" and "spent" cylindrical fuel pin are defined. Calculated results for a spent cylindrical fuel pin are displayed in Fig. 5.5.

The extremum-slope calculated radii are again underestimated as follows:

$$R_{1,\text{ext}} = 4.9453 \text{ mm } (\Delta R_1 = 54.7\ \mu\text{m})$$

$$R_{2,\text{ext}} = 5.9516 \text{ mm } (\Delta R_2 = 48.4\ \mu\text{m})$$

$$R_{3,\text{ext}} = 8.9429 \text{ mm } (\Delta R_3 = 57.1\ \mu\text{m})$$

The consistency of underestimation leads, following the correction suggested in Eq. (5.15), to remarkably accurate estimates of the gap and the cladding thicknesses:

$$\text{Gap}_{\text{est}} = 1.0045 \text{ mm} \qquad (\text{Gap}_{\text{true}} = 1.0000 \text{ mm})$$

$$\text{Cladding}_{\text{est}} = 2.9924 \text{ mm} \qquad (\text{Cladding}_{\text{true}} = 3.0000 \text{ mm})$$

It is clear that the extremum-slope criterion, when corrected, can give extremely accurate dimensional information from radiographs of cylindrical objects.

5.5 INCLUSION SIGNATURES

The distinctions between the converter responses for rectangular and cylindrical objects may potentially be useful in the identification of characteristic signatures of certain inclusions in an object. To suggest this, we consider a cylindrical object containing some cylindrical inclusions, as shown in Fig. 5.6. The object features become particularly pronounced in the first and second derivatives of the converter response function, $\psi'(x)$ and $\psi''(x)$.

The distinction between a cylindrical and rectangular inclusion is suggested in Fig. 5.7. Application of this signature technique places considerable emphasis on the suppression of noise in the optical density of the film.

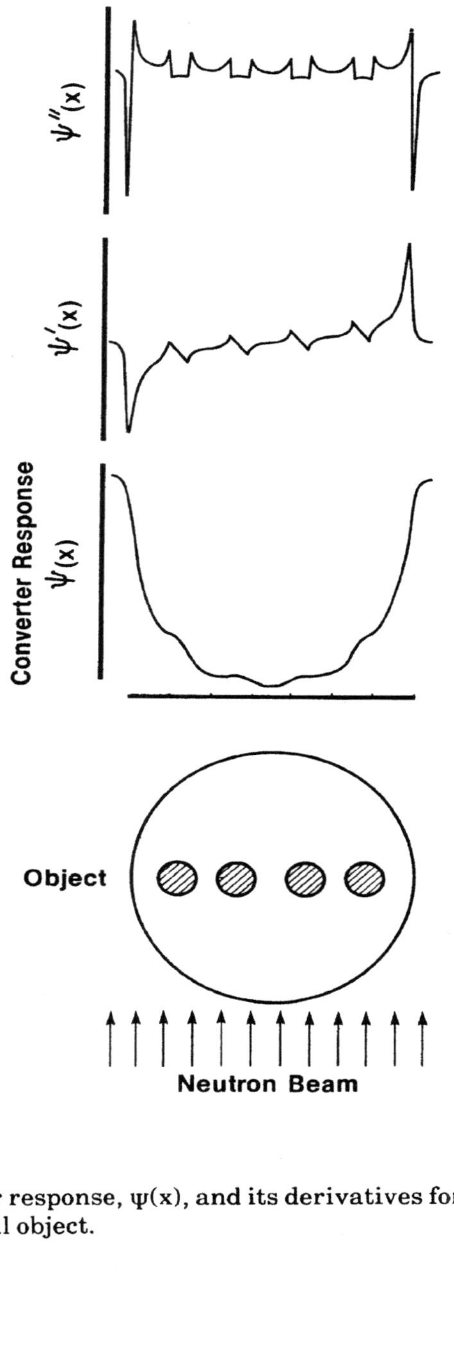

Fig. 5.6: Converter response, ψ(x), and its derivatives for cylindrical inclusions in a cylindrical object.

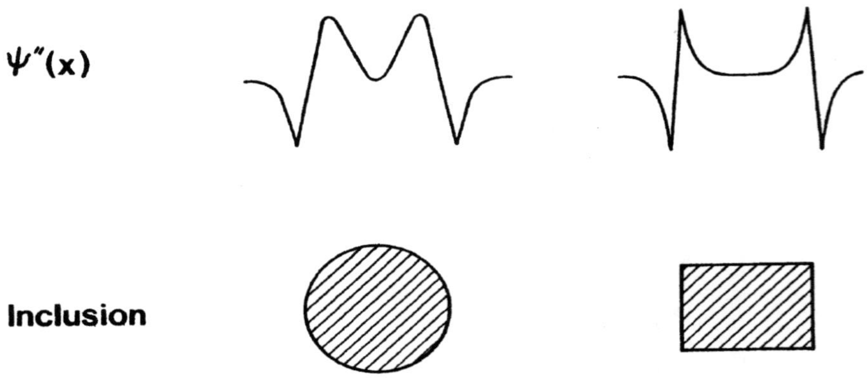

Fig. 5.7: Characteristic variations in the second derivative of the converter response for cylindrical and rectangular inclusions.

VI. Two-Dimensional Analysis

Our previous emphasis to relate an optical image to material details of an object involved principally the analysis of the converter response along a linear traverse of a neutron radiograph. We next extend this one-dimensional analysis to two dimensions, again emphasizing the conceptual-mathematical basis as well as selected applications.

6.1 ONE AND TWO DIMENSIONS

One significant result in Chapters 4 and 5 is the demonstration that a line-spread function,

$$L(x, x_o) = \frac{\lambda/\pi}{1 + \lambda^2 (x - x_o)^2} \tag{6.1}$$

and its various integrants

$$\psi(x) = N \int_{-\infty}^{\infty} \phi_t(x_o)\, L(x, x_o)\, dx_o \tag{6.2}$$

provide useful tools in establishing some precise object-image relationships. The interpretation of the function $L(x, x_0)$ is that it represents the converter response on the image plane at an arbitrary x-coordinate due to an infinitesimally narrow line source of neutron radiation at x_0 in the object plane.

Actually, the coordinate x_0 was defined relative to the objects of interest in Chapters 4 and 5 and not relative to a conceptual object plane, conveniently defined as being orthogonal to the neutron beam and placed adjacent to the object on the exit side of the neutron beam. While this subtle difference was previously unimportant and leaves the previous analysis unaltered, we will now find it simpler to define the position of the neutron sources relative to an object plane, Fig. 1.4. Retaining the convention that the object plane Cartesian coordinates are (u,v), corresponding to (x,y) in the image plane, we can assert that the neutron line source is at the line u_0 in the object plane rather than x_0 in the object. Mathematically, this redefinition is just

$$L(x,x_o) \rightarrow L(x,u_o) = \frac{\lambda/\pi}{1+\lambda^2(x-u_o)^2} \tag{6.3}$$

The line-spread function is one-dimensional since it describes an image plane response with variation in only one of the possible directions, as graphically depicted in Fig. 6.1. Note that the x-y plane and the u-v plane are rigidly related by having their corresponding axes parallel and sharing the same z-coordinate.

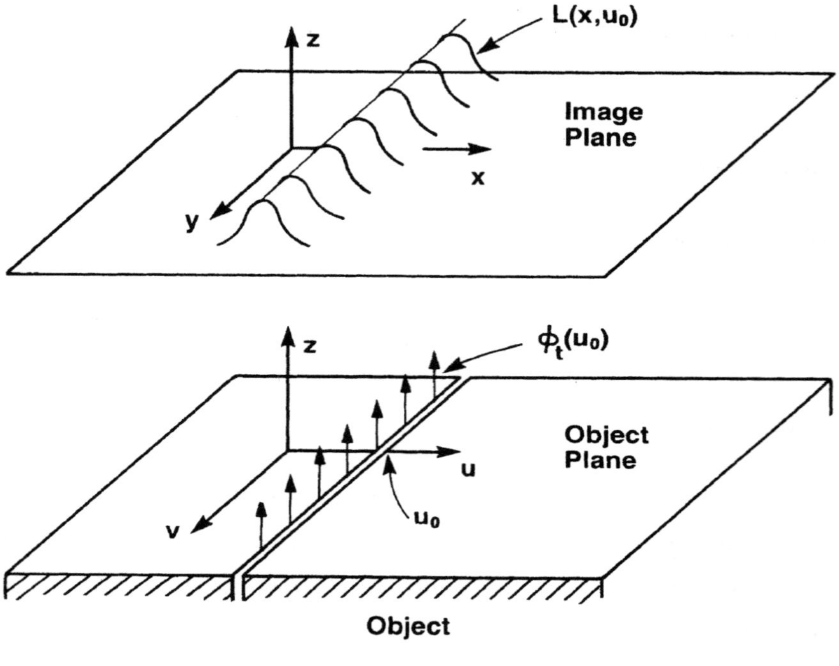

Fig. 6.1: Line-spread function in two dimensions associated with a neutron line source at u_o.

With the depiction of the one-dimensional line-spread function firmly established, it is a self-evident extension to depict a two-dimensional point-spread function, as suggested in Fig. 6.2. The associated point-spread function, $P(x,y,u_o,v_o)$, is therefore the image plane response at an arbitrary point (x,y) in the image plane, attributable to a point source of neutron radiation passing through an infinitesimal hole at the coordinate (u_o,v_o) on the neutron absorbing planar object, or, using equivalent terminology, at (u_o,v_o) in the object plane.

We next consider an explicit expression for, and some properties of, the point-spread function, $P(x, y, u_0, v_0)$.

6.2 POINT-SPREAD FUNCTION

In order to understand the generality of a point-spread function, Fig. 6.2, we note that a line source of transmitted neutron radiation is simply a collection of adjacent point sources arranged linearly. Accordingly, a line-spread function can be constructed by integration of a point-spread function along one object plane dimension (B.7). For example, integrating the point sources from the object plane at $u = u_0$ and along the line $v = v_0$ results in the following image plane converter response at a point (x,y):

$$L(x,y,u_0) = \int_{-\infty}^{\infty} P(x,y,u_0,v_0)\, dv_0 \tag{6.4}$$

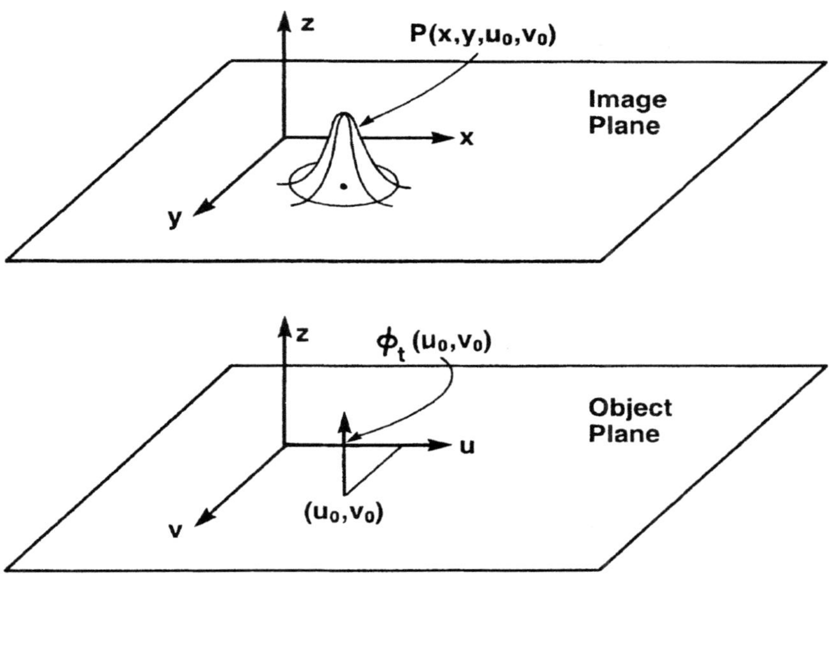

Fig. 6.2: Point-spread function associated with a neutron point source at (u_0,v_0).

Now this response, $L(x,y,u_o)$, will be constant for all y, as indicated by the image plane response constancy in Fig. 6.1. Hence $L(x,y,u_o)$ may simply be abbreviated to $L(x,u_o)$ with the recognition that $L(x,u_o)$ is the image plane converter response at every point (x,y) along the line through x due to a constant neutron source through the line at u_o in the object plane. It is important to recognize that this response constancy will only result if the point source response, $P(x,y,u_o,v_o)$, has the same shape for all source locations (u_o,v_o) in the object plane. Mathematically, this corresponds to the "shift invariance" requirement given by

$$P(x,y,u_o,v_o) = P(x-u_o, y-v_o) \tag{6.5}$$

The integration relation in Eq. (6.4) can now be written as

$$\begin{aligned} L(x,y,u_o) &= \int_{-\infty}^{\infty} P(x-u_o, y-v_o)\, dv_o \\ &= \int_{-\infty}^{\infty} P(x-u_o, y-v_o)\, d(y-v_o) \\ &= L(x-u_o) \end{aligned} \tag{6.6}$$

We thus arrive at the following fundamental relationships between point-spread and line-spread functions for a shift invariant imaging system:

$$L(x-u_o) = \int_{-\infty}^{\infty} P(x-u_o, y-v_o)\, d(y-v_o) \tag{6.7a}$$

$$L(y-v_o) = \int_{-\infty}^{\infty} P(x-u_o, y-v_o)\, d(x-u_o) \tag{6.7b}$$

Additionally, a usually valid geometric property of the point-spread function is circular symmetry, or isotropy, as suggested in Fig. 6.2. This additional property can be shown to imply that

$$P(x,y,u_o,v_o) = P\big((x-u_o)^2 + (y-v_o)^2\big) \tag{6.8}$$

We can now identify the unique point-spread function that integrates, as in Eqs. (6.7), to the familiar line-spread function given in Eq. (6.3) as being

$$P(x-u_o, y-v_o) = \frac{\lambda^2/(2\pi)}{\left\{1 + \lambda^2[(x-u_o)^2 + (y-v_o)^2]\right\}^{3/2}} \tag{6.9}$$

It may be useful to transform the Cartesian (x,y) and (u_o,v_o) coordinates to polar coordinates if this is suggested by the geometry of the object. With reference to Fig. 6.3, we write

$$x = r\sin\theta,\quad u_o = r_o\sin\theta_o$$
$$y = r\cos\theta,\quad v_o = r_o\cos\theta_o \tag{6.10}$$

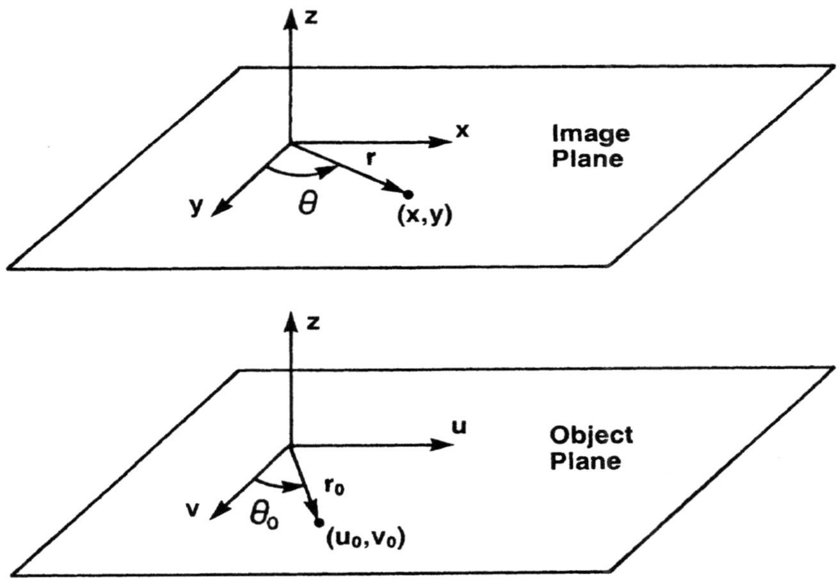

Fig. 6.3: Illustration showing a point on the object plane at coordinate (u_o,v_o) and its corresponding point on the image plane; both Cartesian and polar coordinates are suggested.

Upon substitution in Eq. (6.9), we have, in polar coordinates

$$P(r,\theta,r_o,\theta_o) = \frac{\lambda^2/(2\pi)}{\left\{1 + \lambda^2[r^2 + r_o^2 - 2\,r\,r_o\cos(\theta-\theta_o)]\right\}^{3/2}} \tag{6.11}$$

which may also be compactly written as

$$P(R) = \frac{\lambda^2/(2\pi)}{\{1 + \lambda^2 R^2\}^{3/2}} \tag{6.12}$$

where

$$R^2 = (x - u_o)^2 + (y - v_o)^2 \tag{6.13}$$

Finally, then, by extension of the one-dimensional case, we write for the image plane converter response, resulting from a transmitted neutron beam of intensity $\phi_t(u_o,v_o)$ in the object plane, the two-dimensional integral

$$\psi(x,y) = N \int\int \phi_t(u_o,v_o)\, P(x,y,u_o,v_o)\, du_o\, dv_o \tag{6.14a}$$

Alternatively, in polar coordinates, we have

$$\psi(r,\theta) = N \int\int \phi_t(r_o,\theta_o)\, P(r,\theta,r_o,\theta_o)\, r_o\, dr_o\, d\theta_o \tag{6.14b}$$

Here, N is again a convenient normalization constant. These two expressions are general two-dimensional convolutions and may be written as

$$\psi = N\, \phi_t * P \tag{6.15}$$

where * is the symbol for convolution.

6.3 ABSORBING WEDGE

The application of a point-spread function as a compact and fundamental characterization of two-dimensional radiographic image degradation can be demonstrated through the determination of the image plane response to an absorbing "θ-wedge" object, as shown in Fig. 6.4. With no loss in generality, the neutron absorbing wedge is located with its corner at the origin of the polar coordinate system common to the object and image planes, as defined in Fig. 6.3. For this wedge object,

$$\phi_t(r_o,\theta_o) = \begin{cases} \phi_o, & \theta_o \in (\theta_1, \theta_2) \\ \phi_o \exp[-\Sigma_{tw}\, z_w], & \theta_o \in (\theta_2, 2\pi + \theta_1) \end{cases} \tag{6.16}$$

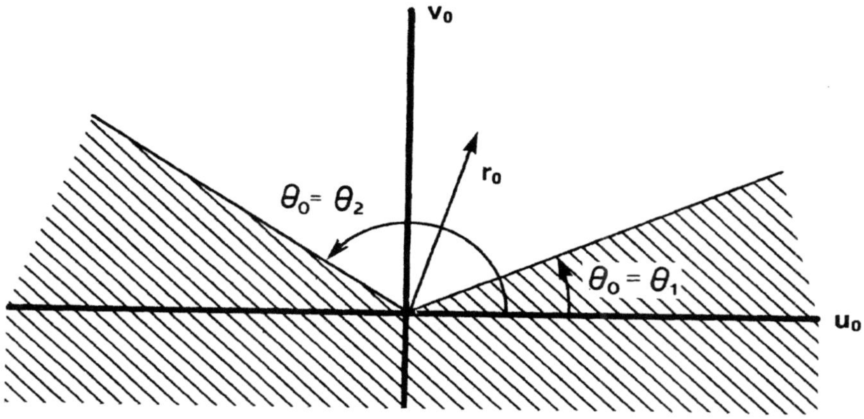

Fig. 6.4: Planar view of an object with a θ-wedge removed.

where Σ_{tw} and z_w are, respectively, the constant absorption cross section and thickness of the solid sector of the wedge.

The converter response is then calculated using Eq. (6.14b) with substitution of Eq. (6.16). The results of the calculation are displayed as radial profiles in Fig. 6.5 for this θ-wedge object in which the wedge angle is $\theta = \pi/2$ and the wedge is neutronically black. The point-spread function in Eq. (6.12) was used for the calculation. Here, the proximity of the radial profiles to step functions is a measure of the two-dimensional unsharpness. These profiles are the two-dimensional analogs of the one-dimensional edge-spread response discussed in Sec. 3.5. Note that for profiles leading into the solid sector of the wedge, unsharpness results in some response in the absence of neutron transmission through the wedge.

It is also useful to note that two angular intervals of directions exist for which the radial converter response profiles increase before decreasing. As these profiles extend from the corner image into the solid sector image, the local object geometry changes from corner to

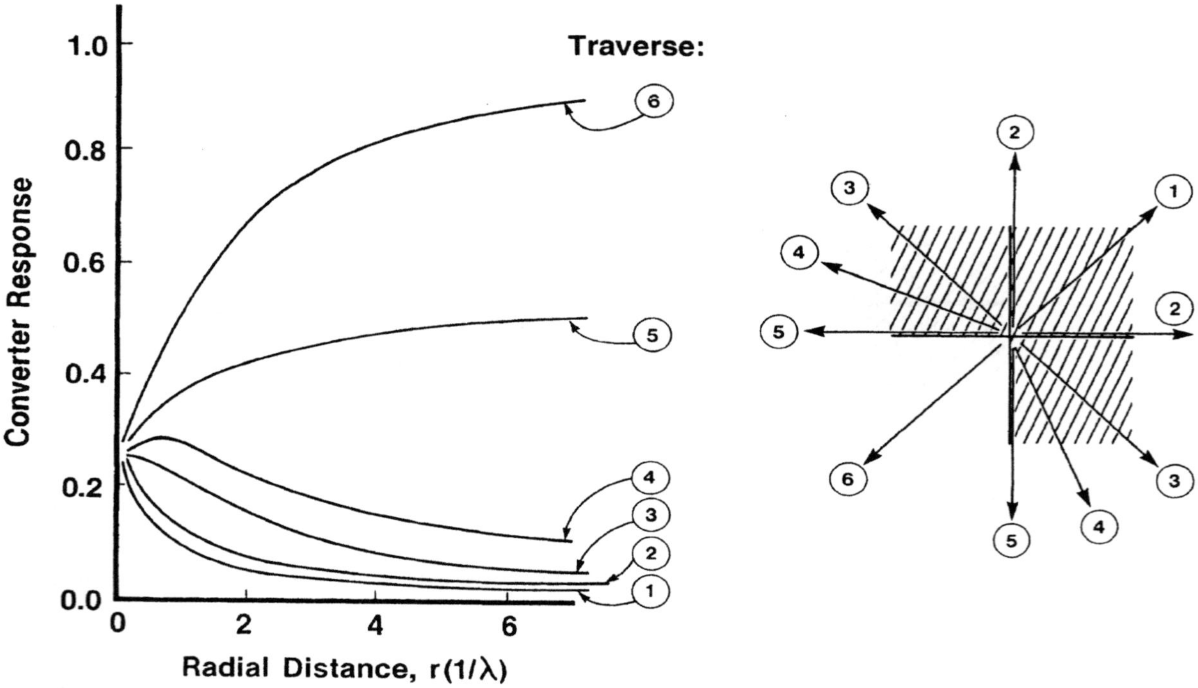

Fig. 6.5: Image plane converter response profiles along the six traverses indicated.

straight edge and finally to a completely filled neutron absorbing plane, corresponding first to an exposure increase and then to the asymptotic decrease towards zero.

The converter response can also be displayed in a contour form in which lines of constant exposure are plotted. A typical exposure contour map is given in Fig. 6.6 for a black $\pi/4$-wedge. In this representation, two-dimensional unsharpness is indicated by, and increases with, contour separation. The contours correspond approximately to the shape of the wedge. In similarity to the radial response profiles, two-dimensional unsharpness is seen to result in a non-zero response to the neutronically black sector of the wedge.

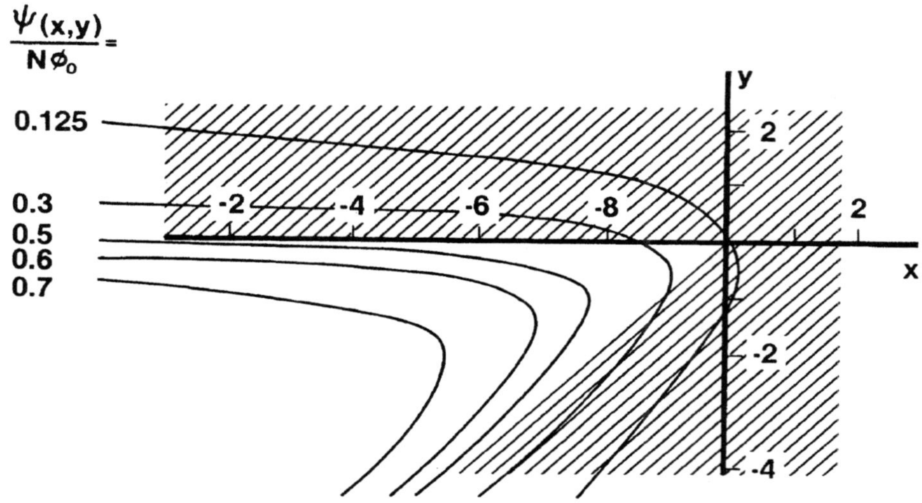

Fig. 6.6: Lines of constant converter response associated with θ-wedge object.

6.4 CORNER LOCATION

An important application of neutron radiography is object dimensioning, as discussed in Chapters 4 and 5, for which the fundamental problem is the location of edges on blurred neutron radiographs. The two-dimensional extension of this one-dimensional problem is the determination of a corner location on a blurred radiograph (A.57). In this context it is understood that 'two-dimensional' refers to the exit neutron beam spatial variation in an object plane. The object itself may possess depth variation and thus be considered as three-dimensional.

To understand the general corner image location problem, we begin with the observation that the converter response at the corner is given by the following evaluation of Eq. (6.14b) in which the polar coordinate origin is at the as yet undetermined corner:

$$\psi_o = \lim_{r \to o} \psi(r_o, \theta_o)$$

$$= \lim_{r \to o} N \int_o^{2\pi} \int_o^{\infty} \phi_t(r_o, \theta_o)\, P\left(r^2 + r_o^2 - 2\, r\, r_o \cos(\theta - \theta_o)\right) r_o\, dr_o\, d\theta_o$$

$$= N \int_o^{2\pi} \int_o^{\infty} \phi_t(r_o, \theta_o)\, P(r_o^2)\, r_o\, dr_o\, d\theta_o \tag{6.17}$$

The differentially attenuated intensity, $\phi_t(r_o,\theta_o)$, is given by

$$\phi_t(r_o,\theta_o) = \phi_o \exp\left[- \int_o^{z_w} \Sigma_t(r_o, \theta_o, z)\right] dz \tag{6.18}$$

where the object depth at (r_o, θ_o) is z_w.

Evaluation of Eq. (6.17) is a difficult if not impossible procedure since it requires a detailed knowledge of the object itself. However, for objects known to possess attenuation properties that are constant along any radius extending from a particular point, such as the θ-wedge of Sec 6.3 or the planar knife-edge, the intensity $\phi_t(r_o,\theta_o)$ reduces to $\phi_t(\theta_o)$ and thus Eq. (6.17) becomes

$$\psi_o = N \int_o^{2\pi} \phi_t(\theta_o)\, d\theta_o \int_o^{\infty} P(r_o^2)\, r_o\, dr_o$$

$$= \frac{N}{2\pi} \int_o^{2\pi} \phi_t(\theta_o)\, d\theta_o$$

$$= N\, \bar{\phi} \tag{6.19}$$

Here, $\bar{\phi}$ is the angle-averaged value of $\phi_t(\theta_o)$. The important implication of Eq. (6.19) is that the corner image plane converter response is determined without the requirement for knowledge of the isotropic system point-spread function.

As an example we again select the θ-wedge of Sec. 6.3 and note that, in practice, the accuracy of this method depends on the degree to which the object's attenuation properties are radially constant about the point of interest. For the θ-wedge with attenuation properties specified by Eq. (6.16), the corner response in Eq. (6.19) becomes

$$\psi_o = N\phi_o \left[\frac{(\theta_2 - \theta_1)}{2\pi} (1 - \exp[-\alpha_w]) + \exp[-\alpha_w] \right] \tag{6.20}$$

where the dimensionless attenuation product is $\alpha_w = \Sigma_{tw} z_w$.

Given a priori knowledge of the attenuation properties of the wedge, Eq. (6.20) implies that the location of the wedge corner can be determined on a highly blurred radiograph by locating a particular point with an optical density, D_o, that corresponds to the converter response, ψ_o. This requires no evaluation of the system point-spread function and no contrast enhancement techniques; the only requirements are local isotropy and shift invariance.

As seen in Fig. 6.6, however, the corner converter response, ψ_o, and therefore its corresponding optical density, D_o, are not isolated. In fact, they lie on contours and the entire image plane can be viewed as a collection of converter response contours. As an example, the corner location for the π/4-wedge in Fig. 6.6 lies on the converter response contour

$$\psi_o = N\,\phi_o \left[\frac{1 + 7\exp[-\alpha_w]}{8} \right] \tag{6.21}$$

This lack of corner response isolation does not preclude the precise location of wedge corners because the corner image is uniquely identifiable as that point about which the corner-valued contour is symmetric. Further, the corner is that point at which the wedge arms intersect, with each wedge arm defined by a line through two or more edge locations. Now the evaluation of edge converter and optical density responses follows immediately from the preceding analysis. The straight edge is simply a special case of the θ-wedge in which the wedge angle is π. From Eq. (6.20), the edge converter response is given by

$$\psi_{edge} = \frac{N\,\phi_o}{2}\,[1 + \mathrm{edge}[-\alpha_w]] \tag{6.22}$$

Note that an intuitive interpretation of Eq. (6.22) is that the edge converter response lies halfway between the maximum and minimum responses, as discussed in Sec. 4.6.

VII. Object Scattering

An examination of neutron radiographic image degradation resulting from object scattered neutrons is undertaken using several approaches of varying complexity and context. Insight into both general and local scattering based imaging effects is the objective.

7.1 SCATTERING DEGRADATION

A frequently important image degradation source in diagnostic radiography is the scattering of the incident probing radiation within the inspected object. While this problem is generally acknowledged in neutron radiography, it is frequently impractical to undertake precise analyses designed to quantify the effects of object scattering in a particular diagnostic setting. The radiographer is more likely to estimate the role of object scattering in general terms such as contrast reduction.

Scattering based image degradation is particularly bothersome since it is impossible to determine which sections of the object have been traversed by the noncollimated scattering portion of the exiting neutron beam. Scattered radiation thus contributes to image degradation, with the most often cited effect being a general reduction in contrast (B.8, B.9). This effect is fundamentally different from the blurring effect induced by converter unsharpness. Rather than spreading existing radiographic information encoded in the exiting beam, or ray image, scattered radiation masks radiographic information through the introduction of extraneous radiation in the ray image. To understand this effect, consider an experiment in which two different regions of an object have images with constant optical densities, D_1 and D_2, such that the first is neutronically the less opaque region and hence $D_1 > D_2$. The object is assumed to be both scattering and absorbing and to some degree of approximation, an approximately constant scattering density, D_s, is a component of both D_1 and D_2. The radiographic recorder is then shifted back from the object by a distance sufficiently great to effectively prevent registry of most of the scattered radiation. The new image will result in local densities of

$$D_1' = D_1 - D_s \tag{7.1a}$$

$$D_2' = D_2 - D_s \tag{7.1b}$$

corresponding to a relative contrast of

$$\frac{D_1' - D_2'}{D_1'} = \frac{D_1 - D_2}{D_1 - D_s} > \frac{D_1 - D_2}{D_1} \tag{7.2}$$

The relative regional contrast is thus smaller with scattered radiation than without. The assumption of spatial constancy of the scattering density component, D_S, is plausible because of the directional divergence of scattered radiation, particularly when thermal neutrons are the probing radiation. We shall see, however, that substantial local scattering effects can occur that actually increase the local image contrast when the scattering component is strongly spatially variable.

Two general classes of methods exist for overcoming the problem of image degradation by scattered radiation. The first of these involves reducing the formation of scattering by restricting the incident beam cross-sectional area through the application of cones, diaphragms and masks. The second class is based on preventing scattered radiation from reaching the recording device. This is accomplished primarily by the insertion of antiscatter grids (A.26, B.8) between the object and the recorder. One problem associated with antiscatter grids is the superposition of the grid pattern onto the resulting image. This problem can be overcome through motion of the grid or by selecting a grid with a strip density sufficiently high that the imaging system possesses a low response to this high spatial frequency. While antiscatter grids should arguably be employed in all neutron radiography, they often are not. A knowledge of object scattering based image degradation is therefore worthwhile.

Given the occurrence of object scattering, it is possible to estimate its extent using certain image quality indicators. These indicators are normally simple test objects that are radiographed together with the object of interest in order to estimate the scattering component in the neutron beam exiting the object. They are simple and yield approximate results. Since the purpose of this chapter is to describe the generation of, and image degradation incurred by, object scattered neutrons, we refer the reader to the complete discussion of neutron image quality indicators found in the literature, e.g. Ref. A.55, B.9.

7.2 COUNTING SCATTERED NEUTRONS

Since it is the uncollided and hence collimated neutron beam exiting an object that forms the information-carrying ray image, we encountered no difficulty in selecting a meaningful neutron population, or intensity, indicator in previous chapters. We have without exception employed the neutron scalar flux, denoted generally by ϕ(n/cm^2-s), as a beam intensity indicator. While it is practical to think of ϕ as counting the neutrons at a point in or outside the object, it should be understood that ϕ is actually a flux density, representing a real and expected number of neutrons passing through some area per unit time only when it is integrated over that area.

These notational considerations correspond well with the reality of measuring a neutron intensity. A real instrument can only estimate the neutron beam intensity at a point by counting neutrons in some neighborhood around that point. If we wish to retain the useful concept of the neutron population at a point, then we must decide specifically how the neutrons are to be counted in that point's neighborhood. The following two counting methods can be employed.

In the first method, all neutrons passing through a unit area centered on the point of interest are counted, over all possible orientations of that area. This mode of counting is useful for calculating neutron-nucleus interaction rates, and the resulting intensity is just the neutron scalar flux, ϕ, employed in past chapters.

In the second method, one counts neutrons passing through a unit area centered on the point of interest, except that only a single orientation of the unit area is considered. Neutrons passing through the area in an arbitrarily defined forward direction comprise the forward partial neutron current, J^+, while their reverse direction counterparts comprise the reverse partial neutron current, J^-. Depending on the circumstances, the neutron intensity can be taken as J^+, J^- or the difference, $J^+ - J^-$, each with units of n/cm^2-s (B.10).

It follows from the definitions that ϕ is always at least as large as J^+, J^- and even $J^+ + J^-$. For example, the neutron population for the situation depicted in Fig. 7.1 is characterized by $J^+ = 20$, $J_- = 10$, $\phi = 30$ n/cm^2-s. Note that for this admittedly contrived case in which the neutrons are perfectly collimated, we have $\phi = J^+ + J^-$. This only holds here because all neutrons in the system can be counted with the single orientation of the unit area. The exact relationship between the scalar flux and partial currents can be established by common reference to a more fundamental quantity, the neutron angular flux, $\psi(\underline{\Omega})$. This symbol is not be confused with $\psi(x,y)$, the converter response of previous chapters. For notational simplicity we will continue to suppress the possible spatial, temporal and energy dependence of J^+, J^-, ϕ and ψ.

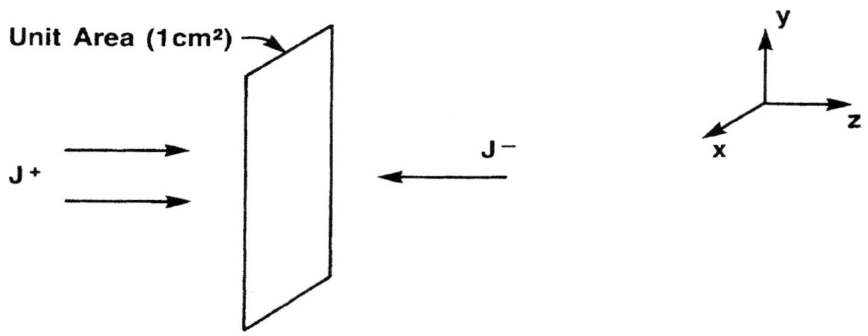

Fig. 7.1: Depiction of J^+ and J^-, with $J^+ = 20$ n/cm²-s and $J^- = 10$ n/cm²-s. The neutron current here is $\underline{J} = (J_x, J_y, J_z) = (0,0,10$ n/cm²-s) and the neutron scalar flux is $\phi = 30$ n/cm²-s.

The angular flux represents the number of neutrons passing through a unit area in a unit time with directions contained in a unit directional cone, $d\underline{\Omega}$. Further, $\psi(\underline{\Omega})$ is also a flux density with units of n/cm²-s-ster. Now the unit area orientation for $\psi(\underline{\Omega})$ is determined by the direction cone, so that counting neutrons over all unit area orientations corresponds to integrating ψ over $\underline{\Omega}$. The neutron angular and scalar flux are therefore related (B.10) by

$$\phi = \int_{4\pi} \psi(\underline{\Omega})\, d\underline{\Omega} \tag{7.3}$$

To connect J^+ and J^- with $\psi(\underline{\Omega})$, we must first define a vector quantity, the neutron current, $\underline{J}$, in the following manner:

$$J = \int_{4\pi} \psi(\underline{\Omega})\, \underline{\Omega}\, d\underline{\Omega} \tag{7.4}$$

To interpret $\underline{J}$ physically, note that its magnitude represents the maximum, over all unit areas centered on a point, of the net number of neutrons passing through the unit area. It relates to $J+$ and $J-$ through the vector inner product

$$J+ - J- = \underline{J} \cdot \underline{e} \tag{7.5}$$

where $\underline{e}$ is a unit vector in the defined forward direction. For example, if $\underline{e}$ is chosen to be along the z-axis, then

$$J^+ = \int_{2\pi^+} \psi(\underline{\Omega})\, \Omega_z \, d\underline{\Omega} \ , \tag{7.6a}$$

$$J^- = \int_{2\pi^-} \psi(\underline{\Omega})\, \Omega_z \, d\underline{\Omega} \ , \tag{7.6b}$$

where Ω_z is the z-component of $\underline{\Omega}$, and $2\pi+$, $2\pi-$ denote the forward and reverse half-planes of directions.

The connection between ϕ, $J+$, $J-$ and $\underline{J}$ is thus established. For counting in neutron radiography, we will find it most convenient to use the quantities ϕ and $J+$. When counting only collimated, and hence uncollided neutrons, ϕ and $J+$ are equivalent. However, when counting directionally varied scattered neutrons, $J+$ is often the more convenient -- and more relevant -- neutron beam intensity indicator.

7.3 SCATTERING TRANSFER FUNCTIONS

Having established several possibilities in Sec. 7.2 for the precise meaning of a neutron beam intensity, we now address the problem of developing a coherent formulation for object scattering with the goal of providing a useful description of the resulting image degradation. A deeper insight into the imaging effects of object scattered neutrons than is afforded by the simple contrast reduction description in Sec. 7.1 can be obtained by casting object scattering into a transfer function framework (A.61) involving the determination of line and point-spread functions.

We begin by examining the traversal of a unit intensity line source of thermal neutrons normally incident on an infinitesimal slab object of thickness Δz, as shown in

Fig. 7.2. Mathematically, it is useful to represent the line source, which itself is a convenient idealization, by a Dirac delta distribution, $\delta(v-\xi_o)$. This "spike" distribution may be thought of as having width approaching zero and height approaching infinity in such a way that the area is exactly unity (B.6). If the slab is nonscattering with absorption cross section Σ_a, then to the Dirac delta input intensity distribution there corresponds an output intensity distribution

$$I_{\Delta z}^{*}(v,\xi_o) = (1 - \Sigma_a \Delta z)\, \delta(v-\xi_o) \tag{7.7}$$

where the asterisk denotes the idealization inherent in the assumption of no scattering. Note that the neutron intensity, which we are denoting generally by "I", is at this point both a scalar flux and a forward partial current.

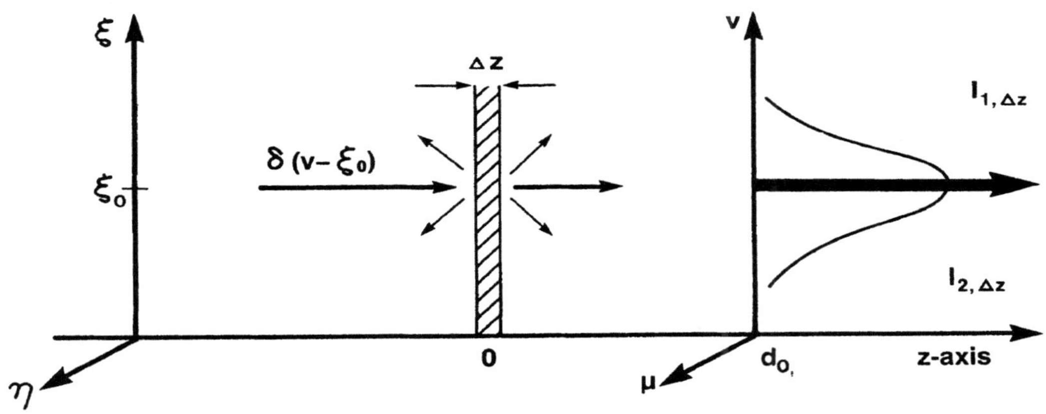

Fig. 7.2: Line source of neutrons traversing Δz and giving rise to a two component intensity distribution at the object plane, $z = d_o$.

To understand the effects of scattering within the slab, it is easiest to view the simple idealized output in Eq. (7.7) as an input to a system, object scattering, which acts to degrade this input. Adding the scattering cross section, Σ_s, to Σ_a to give a total cross section of

$$\Sigma_t = \Sigma_a + \Sigma_s \tag{7.8}$$

suggests that the exiting neutron beam now consists of two components -- the uncollided portion of the beam and the scattering contribution. Note that multiple interactions can be ignored due to the infinitesimal slab width. The uncollided beam component has an intensity distribution given by

$$I_{1,\Delta z}(v,\xi_o) = (1 - \Sigma_t \Delta z)\,\delta(v - \xi_o) \tag{7.9}$$

For isotropic scattering, the scattering component intensity distribution can be written as

$$I_{2,\Delta z}(v,\xi_o) = \left(\frac{\Sigma_s \Delta z}{2}\right) L(v,\xi_o) \tag{7.10}$$

The factor of 1/2 in Eq. (7.10) arises because only half of the scattering is forward. Also, $L(v, \xi_o)$ is the line-spread function in Eq. (3.17) with λ as the inverse separation of the slab and the object plane located at $z = d_o$. An examination of the derivation for L( ) in Sec. 3.3 reveals that this distribution arises from the combination of parallel planes and isotropic emission, such as indicated in Fig. 7.2.

The exiting two component intensity distribution at the plane $x = d_o$ is therefore

$$I_{\Delta z}(v,\xi_o) = I_{1,\Delta z}(v,\xi_o) + I_{2,\Delta z}(v,\xi_o)$$

$$= (1 - \Sigma_t \Delta z)\,\delta(v - \xi_o) + \frac{\Sigma_s \Delta z}{2\pi d_o}\left[1 + \left(\frac{v - \xi_o}{d_o}\right)^2\right]^{-1} \tag{7.11}$$

At this point the exact interpretation of intensity becomes fixed. Since the scattering intensity component was derived by counting neutrons traversing a fixed, vertical plane, the intensities are therefore forward partial currents. This vertical plane is an object plane since it is used as a location at which the ray image intensity is assessed, prior to conversion into secondary radiation.

In the context of object scattering, Eq. (7.7) can be viewed as representing an "input" which, when operated upon by the physical process of object scattering, yields the "output" given in Eq. (7.11). Normalizing to an input of unit integrated intensity, the output is then the desired scattering line-spread function, $L_s(\,)$, given by

$$L_s(v,\xi_o) = \frac{1}{(1 - \Sigma_a \Delta z)}\left\{(1 - \Sigma_t \Delta z)\,\delta(v - \xi_o) + \frac{\Sigma_s \Delta z}{2\pi d_o}\left[1 + \left(\frac{v - \xi_o}{d_o}\right)^2\right]^{-1}\right\} \tag{7.12}$$

This input, output and scattering system relationship is shown schematically in the top diagram of Fig. 7.3.

While the input that generates $L_s(\ )$ is normalized to unity, the output, which is $L_s(\ )$ itself, has an area of

$$\int_{-\infty}^{\infty} L_s(v, \xi_o)\, dv = 1 - \left(\frac{\Sigma_s \Delta z}{2}\right)(1 - \Sigma_a \Delta z)^{-1} \tag{7.13}$$

which is smaller than unity. This arises because the transition from input to output involves backscattering losses.

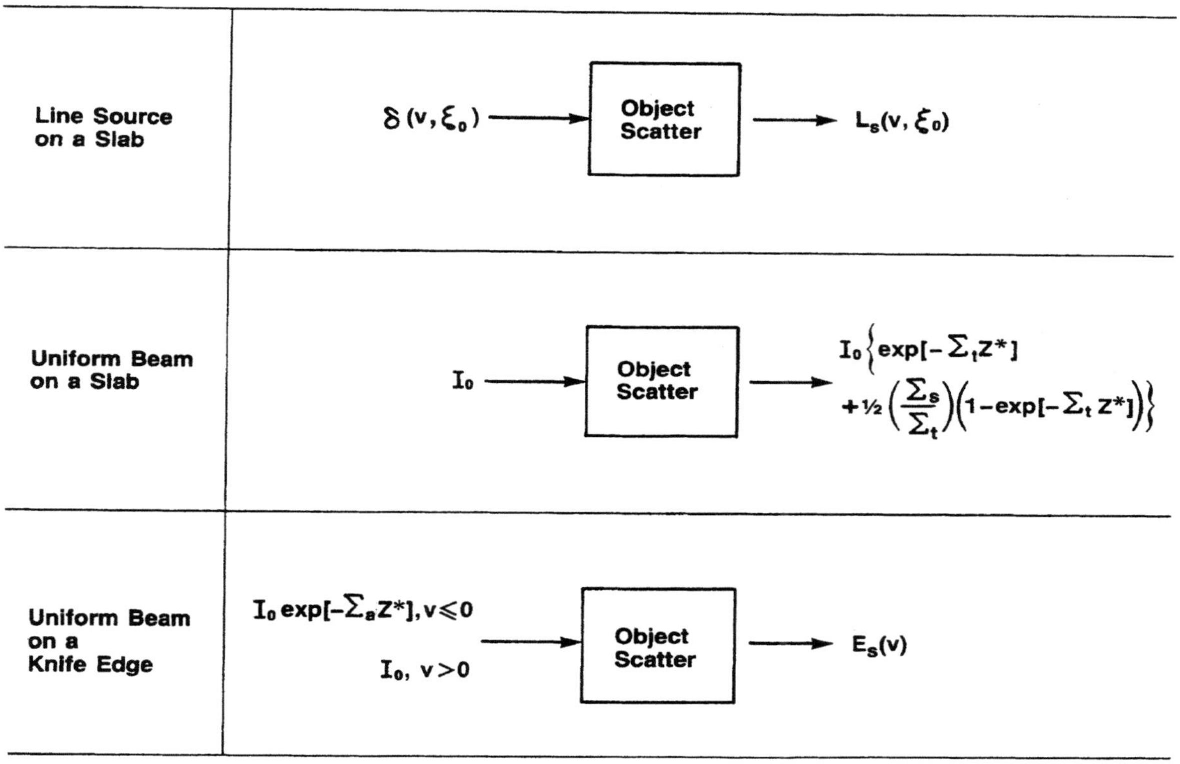

Fig. 7.3: Block-diagram of object scattering as an image degradation process.

The above model can be enhanced at the expense of simplicity. For example, by expanding the slab from infinitesimal width, Δz, to finite width, z^*, while retaining single scattering as an approximation, the resulting line-spread function is given by

$$L_s(v,\xi_o) = \exp[-\Sigma_s z^*]\delta(v-\xi_o)$$

$$+\left(\frac{\Sigma_s}{2\pi}\right)\exp[\Sigma_a z^*]\int_o^{z^*}\left\{[d_o - z]\left[1+\left(\frac{v-\xi_o}{d_o - z}\right)^2\right]\right\}^{-1}\exp[-\Sigma_t z]\,dz \tag{7.14}$$

In light of the discussion on point-spread functions in Sec. 6.2, we can immediately identify the isotropic scattering point-spread function for a point source of neutrons normally incident on this slab of width z*. This point-spread function, which corresponds to the line-spread function in Eq. (7.14), is given by

$$P_s(R) = \exp[-\Sigma_s z^*]\delta(R)$$

$$+\left(\frac{\Sigma_s}{4\pi}\right)\exp[\Sigma_a z^*]\int_o^{z^*}(d_o - z)^{-2}\left[1+\left(\frac{R}{d_o - z}\right)^2\right]^{-3/2}\exp[-\Sigma_t z]\,dz \tag{7.15}$$

where R denotes distance from the point of emission.

The line and point-spread functions discussed above are examples of ways of mathematically transforming an input to an output, and hence are referred to as system transfer functions. Experimental determinations of object scattering point-spread functions have also been undertaken (A.56). While the particular transfer functions derived here are restricted by the assumption of single interactions in an infinite slab, considerably greater flexibility is provided by the transfer function framework itself. This framework is limited only by the accuracy of the transfer functions employed and enables a combination of object scattering effects with other forms of image degradation.

7.4 INFINITE SLAB RESPONSE

The application of scattering transfer functions can be illustrated through the determination of the object plane intensity resulting from a collimated neutron beam of uniform intensity I_0 n/cm^2-s normally incident on an effectively infinite slab, Fig. 7.4. In this case the input, defined as the collimated beam output intensity in the absence of scattering, is

$$I(v) = I_0 \exp[-\Sigma_a z^*] \tag{7.16}$$

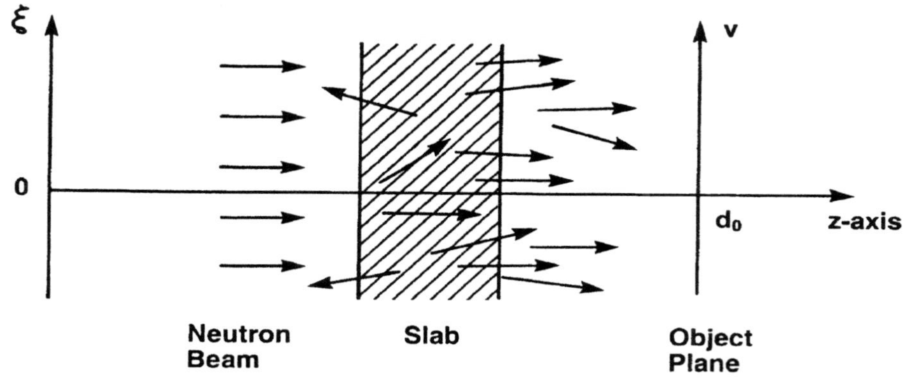

Fig. 7.4: Neutron scattering and absorption in a slab object.

The scattering system output is then the convolution, Sec. 3.6, of I(v) with a scattering line-spread function, Eq. (7.14):

$$O(v) = \int_{-\infty}^{\infty} I(v_o)\, L_s(v, v_o)\, dv_o \tag{7.17}$$

where v_0 is a dummy variable for v.

Upon integration and simplification, Eq. (7.17) reduces to the constant value

$$O = I_o \exp[-\Sigma_t z^*] + \frac{I_o}{2}\left(\frac{\Sigma_s}{\Sigma_t}\right)(1 - \exp[-\Sigma_t z^*]) \tag{7.18}$$

The first term represents the uncollided exiting neutron beam while the second term represents that portion of the collided beam that is scattered forward. Both terms represent the neutron beam prior to conversion into secondary radiation. The role of object scattering in the infinite slab response is suggested schematically in the second diagram of Fig. 7.3.

7.5 KNIFE-EDGE RESPONSE

One effect of converter unsharpness in neutron radiography is the blurring of edge images. This blurring was modelled as an edge-spread function in Sec. 3.6. The effects of object scattering on edge images can be calculated using a scattering line-spread function, Sec. 7.3. To this end, we employ as an object the knife-edge slab of Fig. 3.6., reproduced with relevant notation in Fig. 7.5. The input intensity here is

$$I(v) = \begin{cases} I_o \exp[-\Sigma_a z^*], & v \leq 0 \\ I_o, & v > 0 \end{cases}, \tag{7.19}$$

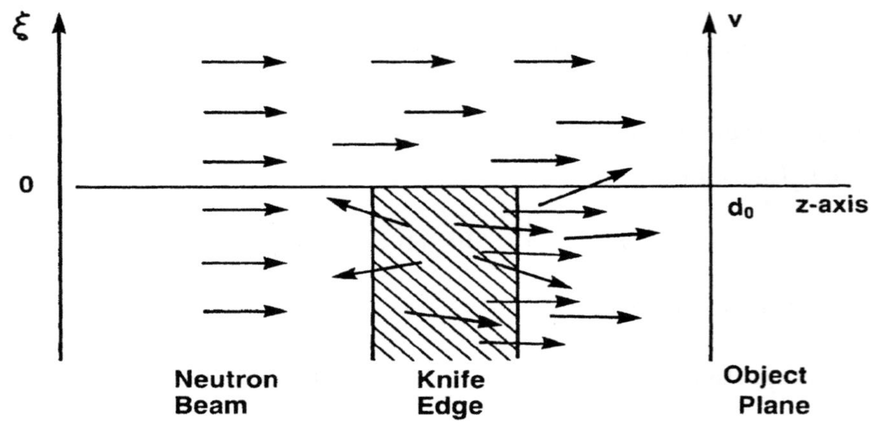

Fig. 7.5: Neutron beam interaction in a knife-edge slab object.

When passed through the object scattering image degradation system, represented mathematically by L( ) in Eq. (7.14), this input is transformed to the output edge response function given by the following convolution:

$$E_s(v) = \int_{-\infty}^{\infty} I(v_o)\, L_s(v, v_o)\, dv_o$$

$$= I_o \left[H(v) + \exp[-\Sigma_t z^*]\, H(-v) + \frac{I_o}{2}\left(\frac{\Sigma_s}{\Sigma_t}\right) \mathrm{ftan}\left(\frac{v}{d_o}\right)\right] \tag{7.20}$$

Here, H(v) is the Heaviside step function, defined (B.6) by

$$H(v) = \begin{cases} 1, & v > 0 \\ 0, & v < 0 \end{cases}, \tag{7.21}$$

and f tan( ) is defined by

$$\mathrm{ftan}(v) = \frac{1}{2} - \left(\frac{1}{\pi}\right) \tan^{-1}(v) \tag{7.22}$$

The effect of object scattering on an edge response is included in Fig. 7.4 while the scattering edge response, $E_s(v)$, is shown in Fig. 7.6. The distinguishing features are the two inverted but otherwise identical hump-shaped distortions on either side of the knife-edge location, centered on, and diminishing in magnitude with, distance from the edge location. This scattering edge response represents the neutron beam intensity, a forward partial current, prior to conversion into secondary radiation.

As suggested by the included response curve for the case of no scattering, these distortions are strictly a result of object scattering (A.62). Quite simply, neutrons scattering at $\xi < 0$, Fig. 7.5, will fractionally enter the $\xi > 0$ region but the reverse does not occur. This results in a net outward current from the solid portion of the slab in the ξ-direction so that, relative to the nonscattering response, there is an increase in intensity for $v > 0$ and a decrease for $v < 0$. The relative decrease also has a constant component due to backscattered neutrons. This explains the existence of the scattering based edge distortions. To explain their shape, we recognize that localized edge effects must vanish at large distances from the edge so that the distortions should diminish with distance from the edge. Further, the distortions elongate with increasing separation of the selected object plane from the slab.

The radiographic interpretation of these distortions is a visual accentuation of edges. Thus, while object scattering can induce a general reduction in contrast, it can simultaneously result in local contrast enhancement. It follows that such enhancement occurs at

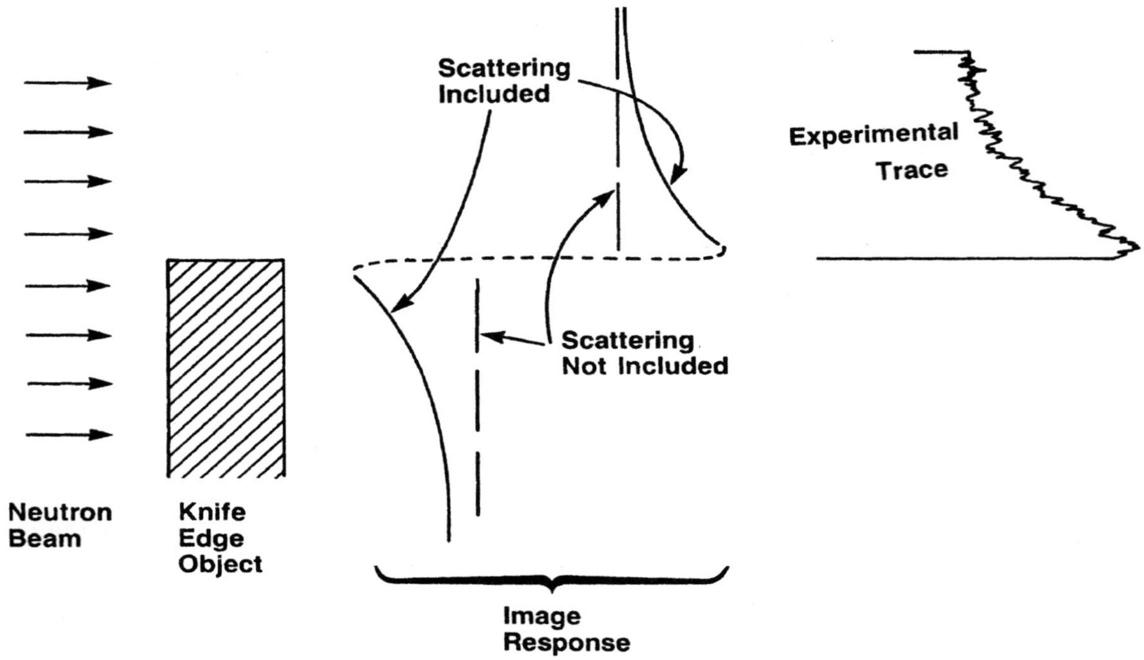

Fig. 7.6: Effect of neutron scattering in a knife-edge slab object.

any material interface or boundary, since these are regions where an unbalanced scattering flux exists. Since, however, the distortions elongate with increasing separation of the object plane and the object, the accentuation of edges is effectively removed by placing the neutron converter a few centimeters away from the object. The separation distance required to suppress this artifact is approximately proportional to Σ_s itself.

The edge response model developed in this section can be made more realistic if the effects of converter unsharpness are incorporated. As we will see in the next chapter, this can be accomplished by further convolution of the edge response, Eq. (7.20), with a line-spread function representing converter unsharpness.

We digress briefly to observe that image edge accentuation commonly results from a variety of other sources as well. For example, the application of common edge enhancement filters to blurred edges, such as those indicated by Fig. 3.9, results in a very similar effect referred to as "ringing" or "overshoot" (B.11). Another effect, resulting from the spatial diffusion of developer during the film development process, also produces these edge distortions. Each of these edge enhancement effects is compounded with a natural edge enhancement performed in normal vision.

7.6 NEUTRON BUILD-UP FACTOR

Having formally established neutron beam intensity indicators relevant to neutron radiography in Sec. 7.2, we can proceed with an equation based approach towards estimating the magnitude of object scattering in neutron radiography. The relevant governing equation here is the neutron transport equation which, in its most general form, can appear as either an integral or integro-differential equation (B.12). Since a prohibitively large amount of work is involved in solving this equation in its general form, approximations are invariably made to suit the particular material configuration of interest. For beam intensity determinations in neutron radiography this approach has less appeal than the transfer function method in which approximate line or point-spread functions are linearly combined to yield an output beam intensity. A combination of these two approaches would involve obtaining transport equation solutions to determine the transfer functions.

As an example of equation based scattering calculations of particular relevance to neutron radiography, we present here a transport equation solution for the infinite slab examined in Sec. 7.4. The geometry is reproduced with appropriate notation in Fig. 7.7.

At any point inside or outside the object the intensity can be written as the sum of an uncollided term, I*, and a scattering term, as follows:

$$I(\underline{P}) = I^*(\underline{P}) + \int_{\text{all } P'} I(\underline{P}')\,\Sigma_s(\underline{P}')\,K(\underline{P},\underline{P}')\,d\underline{P}' \tag{7.23}$$

Here, $I^*(\underline{P})$ is given, Sec. 2.4, by:

$$I^*(P) = \begin{cases} I_o, \; z \le 0 \\ I_o \exp\left[-\int_0^z \Sigma_t(u,v,w)\,dw\right], \; z \in [0, z^*] \\ I_o \exp\left[-\int_0^{z^*} \Sigma_t(u,v,w)\,dw\right], \; z \ge z^* \end{cases} \tag{7.24}$$

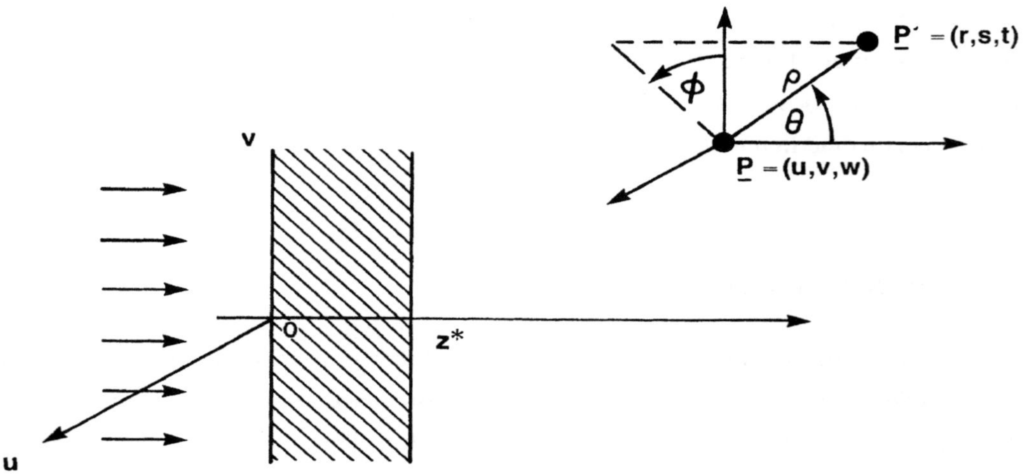

Fig. 7.7: Object slab geometry and polar coordinates.

where w is a dummy variable for z. The second term in Eq. (7.24) describes the intensity contribution from neutrons at all other points $\underline{P}' = (r,s,t)$. The point scattering kernel, $K(\underline{P}, \underline{P}')$, is the probability that a neutron scattered at the point $\underline{P}'$ travels without further interaction to the point $\underline{P}$. Equation (7.23) is a Fredholm integral equation of the second kind (B.6) and is essentially a specialization of the general integral neutron transport equation. For isotropic scattering, such as is typically valid for the scattering of thermal neutrons in a nonhydrogenous medium, $K(\underline{P}, \underline{P}')$ is the following combination of exponential attenuation and inverse square spherical divergence:

$$K(\underline{P},\underline{P}') = \frac{\exp\left[-\Sigma_t(\underline{P}')\,\alpha(\underline{P},\underline{P}')\,|\underline{P}-\underline{P}'|\right]}{4\pi|\underline{P}-\underline{P}'|^2} \tag{7.25}$$

Here, $\alpha(\ )$ is a geometric variable that is simply unity when $\underline{P}$ is inside the object, and is the following ratio of z coordinates for $\underline{P}$ outside the object:

$$\alpha(\underline{P},\underline{P}') = \begin{cases} \left|\dfrac{t}{z-t}\right|, & z \le 0 \\ \left|\dfrac{z^*-t}{z-t}\right|, & z \ge z^* \end{cases} \tag{7.26}$$

For a homogeneous slab, it is expected that

$$I(\underline{P}) = I(z) \tag{7.27}$$

so that the governing equation, Eq. (7.23), reduces to

$$I(z) = I^*(z) + \Sigma_s \int_0^{z^*} I(t)\, f(t, z)\, dt \tag{7.28}$$

where

$$f(t,z) = \int_{-\infty}^{\infty} \int_{-\infty}^{\infty} K(r-u, s-v, t-z, \alpha)\, d(r-u)\, d(s-v) \tag{7.29}$$

It is easiest to solve Eq. (7.28) by switching to spherical polar coordinates, defined in Fig. 7.8. Equation (7.28) then transforms for $z \in [0,z^*]$ to

$$I(z) = I_o \exp[-\Sigma_t z] + \frac{\Sigma_s}{2} \left\{ \int_0^1 \int_0^{(z^*-z)/\mu} I(z+\rho\mu) \exp[-\Sigma_t \rho]\, d\rho\, d\mu + \int_{-1}^{0} \int_0^{-z/\mu} I(z+\rho\mu) \exp[-\Sigma_t \rho]\, d\rho\, d\mu \right\} \tag{7.30}$$

where $\mu = \cos(\theta)$. For $z \geq z^*$ and $z \leq 0$, the intensities are constant at $I(z^*)$ and $I(0)$, respectively, since there are no neutron removal processes outside the slab.

Thus, even for this simple geometry, the solution to another formidable integral equation is required. However, Eq. (7.30) has an easily computable, although not analytical, Neumann series solution (B.6). This iterative solution is initiated by substituting the initial estimate

$$I^{(o)}(z) = I_o \exp[-\Sigma_t z] \tag{7.31}$$

into the right hand side of Eq. (7.30), which is then evaluated to yield the next estimate, $I^{(1)}(z)$, on the left hand side. This estimate is similarly used to generate $I^{(2)}(z)$, and so on. The complete solution can be expressed in series form as

$$I(z) = I_o \exp[-\Sigma_t z] \sum_{i=0}^{\infty} \left(\frac{\Sigma_s}{2}\right)^i f_i(z) \tag{7.32}$$

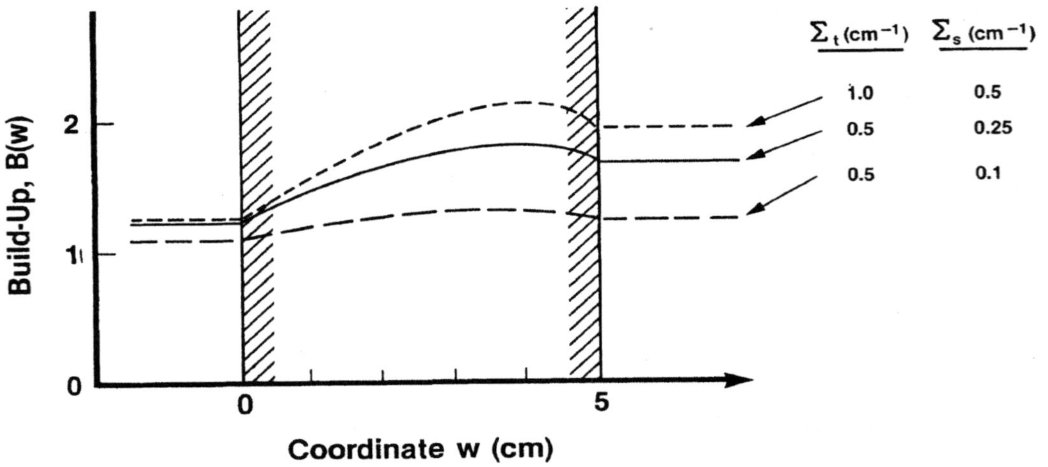

Fig. 7.8: Neutron build-up factor in a slab for various cross section combinations.

where the f_i modes obey the recurrence relation

$$f_i = \int_{-1}^{o} \int_{o}^{-z/\mu} \exp[-\rho\, \Sigma_t (1+\mu)]\, f_{i-1}\, d\rho\, d\mu$$

$$+ \int_{o}^{1} \int_{o}^{(z^* - z)/\mu} \exp[-\rho\, \Sigma_t (1+\mu)]\, f_{i-1}\, d\rho\, d\mu \qquad (7.33)$$

with $f_o = 1$.

A useful property of this solution is that the i'th term represents neutrons that have undergone precisely i scatterings. The scattering kernel, Eq. (7.25), was formulated to yield intensities, I(z), that are actually neutron scalar flux. By introducing a factor of μ into the integrands in Eq. (7.33), the resulting intensities become forward partial currents, in which case the first mode in Eq. (7.32) becomes the infinite slab response developed in Sec. 7.4.

The solution, Eq. (7.32), can also be written as

$$I(z) = I_o \exp[-\Sigma_t z]\, B(z) \tag{7.34}$$

where

$$B(z) = \sum_{i=0}^{\infty} \left(\frac{\Sigma_s}{2} \right)^i f_i \tag{7.35}$$

Here, B(z) is the build-up factor, Sec. 2.4, a function defined as the ratio of total (collimated plus scattered) to scattered neutron intensity. All information regarding neutron scattering is contained in B(z), several examples of which are plotted in Fig. 7.8. The reader should be alerted to other definitions of "build-up" such as, for example, that used in megavoltage radiotherapy physics (B.13).

This model is of particular relevance to neutron radiography since the equation solution provides, in B(z), a precise indication of the relative scattering content in the neutron beam exiting the object and directed towards the neutron converter. For example, it follows from Fig. 7.9 that for a slab-like object with minor material variations and with cross sections in the range $\Sigma_t \simeq 1.0\,cm^{-1}$, $\Sigma_s \simeq 0.5\,cm^{-1}$, the exit build-up factor is approximately 2.0. This implies that roughly half of the exiting neutron beam consists of scattered neutrons, resulting in a substantial decrease in contrast, Sec. 7.1. If a solution is calculated based on an intensity defined as the forward partial neutron current, then the build-up factor is considerably smaller, as illustrated in Fig. 7.9.

Several points regarding the scope of this model should be mentioned here. First, we note that this model includes not only multiple scattering, but also the absorption of multiple scattered neutrons. Second, this model can be extended to calculate a scattering line-spread function for more general application in a linear systems framework, if we specify a line source boundary condition rather than the uniform plane source in Fig. 7.7. However, this would greatly complicate the solution procedure. For example, if a horizontal line source were employed, then a solution for I(v,z) would be required as opposed to I(z) for a plane source.

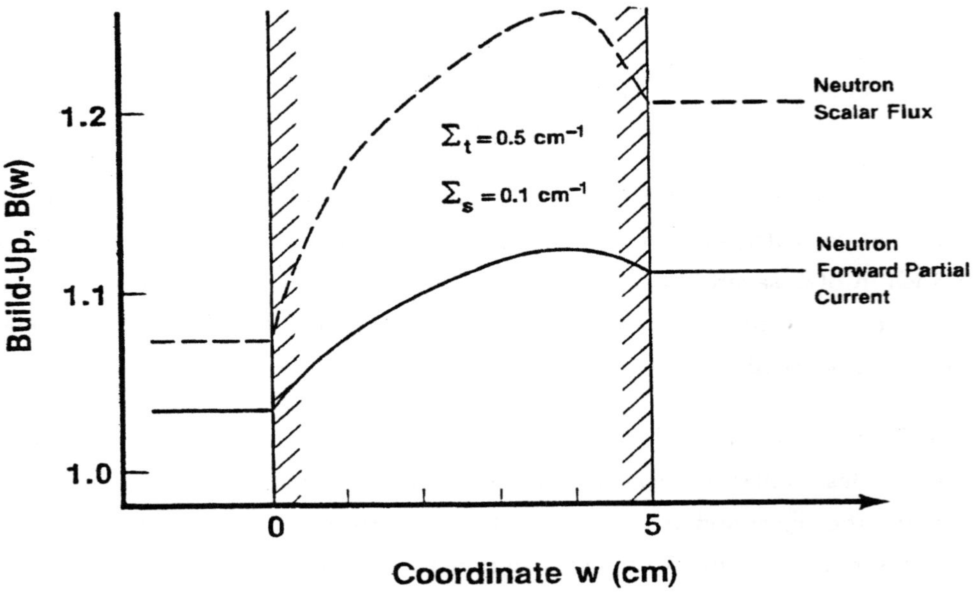

Fig. 7.9: Illustration of the difference between the neutron scalar flux and the neutron forward partial current build-up factors in a slab.

A good method for calculating scattering transfer functions and allowing multiple scattering is analog Monte Carlo simulation (A.62, B.14). This method involves the simulation of particle collision trajectories, or histories, using random numbers. While it can demand long computation time, an analog Monte Carlo formulation is usually, for a given accuracy of solution, much more manageable than equation-based approaches. The application of Monte Carlo methods is not within the scope of this text and will thus be discussed no further.

VIII. Linear Systems Formulation

Selected aspects of linear systems theory have been employed in previous chapters within the context of the convolution integral. We explore this approach further and illustrate it by specific applications.

8.1 TRANSFER FUNCTIONS AND SYSTEM LINEARITY

A fundamental premise in linear systems analysis is that a system's output response for a general input can be expressed as the summation of output responses to simple inputs. This is also referred to as the principle of superposition. In neutron radiography the "system" can be taken as any individual image degradation component, such as converter unsharpness, or any combinations thereof. The system input is invariably a radiation intensity, either neutrons or converter-generated secondary radiation, depending on the system being analyzed.

We have seen that useful simple inputs, from which a general input can be constructed, are line and point sources. The system output responses are referred to as line and point-spread functions, L( ) and P( ). These are examples of system transfer functions. The usefulness of line and point sources stems from the fact that the function value of a general input function, I( ), at any point or line is precisely the magnitude of the point or line source comprising the input at that location. This leads to convolution expressions like

$$O(x) = \int_{-\infty}^{\infty} I(u)\, L(x,u)\, du \tag{8.1}$$

$$O(x,y) = \int_{-\infty}^{\infty} I(u,v)\, P(x,y,u,v)\, du\, dv \tag{8.2}$$

where the system output is O( ), x and y are Cartesian coordinates, and the system is modelled as being noise-free. The system is referred to as linear because the output for any weighted sum of inputs equals the same weighted sum of outputs to the individual inputs. For example, Eq. (8.1) describes the output, O( ), to an input, I( ), and so we can write O = O(I). Define an input, I_3, as a weighted sum of two other inputs, I_1 and I_2, as follows:

$$I_3(u) = \alpha I_1(u) + \beta I_2(u) \tag{8.3}$$

Then the output to I_3 ( ) is

$$\begin{aligned} O(I_3) &= \int_{-\infty}^{\infty} [\alpha I_1(u) + \beta I_2(u)]\, L(x,u)\, du \\ &= \alpha \int_{-\infty}^{\infty} I_1(u)\, L(x,u)\, du + \beta \int_{-\infty}^{\infty} I_2(u)\, L(x,u)\, du \\ &= \alpha\, O(I_1) + \beta\, O(I_2) \end{aligned} \tag{8.4}$$

and the system is therefore linear. Inspection of Eq. (8.4) reveals that linearity results when the transfer function, L( ) or P( ), exhibits no dependence on the intensity itself.

Image degradation sources, such as converter unsharpness, can usually be properly modelled as a linear system. Object scattering,. however, is an inherently nonlinear image degradation source. In the general case, this nonlinearity arises because the scattering point-spread function varies spatially according to the object's scattering properties and thus depends on the object itself (B.4). Mathematically, this corresponds to

$$O(x,y) = \int_{-\infty}^{\infty}\int_{-\infty}^{\infty} I(u,v)\, P(x,y,u,v,\, I(u,v))\, du\, dv \tag{8.5}$$

so that using the notation O(I), with $I_3 = \alpha I_1 + \beta I_2$, we have

$$\begin{aligned} O(I_3) &= \int_{-\infty}^{\infty}\int_{-\infty}^{\infty} (\alpha I_1 + \beta I_2)\, P(x,y,u,v,I_3)\, du\, dv \\ &\neq \alpha \int_{-\infty}^{\infty}\int_{-\infty}^{\infty} I_1\, P(x,y,u,v,I_1)\, du\, dv + \beta \int_{-\infty}^{\infty}\int_{-\infty}^{\infty} I_2(x,y,u,v,I_2)\, du\, dv \end{aligned} \tag{8.6}$$

This nonlinearity can be removed if P( ) can be approximated without precise knowledge of the object, or equivalently, I( ). If the goal of the analysis is to calculate an output from an input, the problems posed by this nonlinearity can in principle be overcome, since P( ) can be found given I( ). This was the approach adopted in Ch. VII. The nonlinearity problem can be much more difficult if the goal is to reconstruct the input given an output since, in this case, there may be no direct way to calculate P( ). The inversion, or deconvolution, of Eqs. (8.1) and (8.2) to determine the original undegraded input, I( ), is an example of an image restoration problem.

8.2 FOURIER TRANSFORM

The application of linear systems theory in any form of imaging analysis frequently involves a powerful tool known as the Fourier transform. For functions obeying the Dirichlet conditions -- and this includes most functions of interest in applied physics -- it is possible to decompose the function into sinusoids of varying frequency. The Fourier transform represents the limit of such a decomposition in which the function is expressed as a continuum of magnitudes over all frequencies. For imaging problems the frequency is usually a spatial frequency with units of inverse length. The Fourier transform can thus be thought of as a tool that converts a function of spatial dependence to a function of spatial frequency dependence, analogous to some phenomenon being described in two different languages. For one and two-dimensional functions, f(x) and g(x,y), the complex Fourier transform is defined (B.15) as

$$\mathcal{F}\{f(x)\} = F(v_x) = \int_{-\infty}^{\infty} f(x)\exp[-2\pi i x\, v_x]\,dx \tag{8.7}$$

$$\mathcal{F}\{g(x,y)\} = F(v_x,v_y) = \int_{-\infty}^{\infty}\int_{-\infty}^{\infty} g(x,y)\exp[-2\pi i\,(x\, v_x + y\, v_y)]\,dx\,dy \tag{8.8}$$

To illustrate an application of the Fourier transform in neutron radiographic and other imaging analysis, we extend the two-dimensional imaging equation, Eq. (8.2), to the following input-output form:

$$O(x,y) = N\int_{-\infty}^{\infty}\int_{-\infty}^{\infty} I(u,v)\,P(x,y,u,v)\,du\,dv + n(x,y) \tag{8.9}$$

In this equation an imaging system is represented mathematically by the transformation of an input, I( ), to an output, O( ), through the system point-spread function, P( ). The system noise component in the output is denoted by n( ). For a shift invariant system, recall, Sec. 6.2, that

$$P(x,y,u,v) = P(x-u, y-v) \tag{8.10}$$

In this case, Eq. (8.9) can be written using convolution notation as

$$O(x,y) = N\, I(x,y) * P(x,y) + n(x,y) \tag{8.11}$$

Typically, O( ) and P( ) are obtained through measurement. For example, O( ) might be a spatial exposure or optical density variation whereas P( ) might be obtained from the line-spread function, Sec. 6.2. The task of the analyst will often be to unfold, or "deconvolve", the original object plane input intensity distribution, I( ), from the measured output, O( ), given the point-spread function, P( ). This deconvolution can be performed in the spatial frequency domain, using the Fourier transform and the convolution theorem, which states that the Fourier transform of the convolution of two functions is the product of their Fourier transforms (B.15). Application of the Fourier transform to Eq. (8.11) then results in

$$O(v_x, v_y) = N\, I(v_x, v_y)\, P(v_x,v_y) + n\,(v_x, v_y) \tag{8.12}$$

Here, v_x and v_y are spatial frequencies in the x and y directions, while O( ), I( ) and n( ) are complex spatial frequency spectra. The transformed point-spread function, $P(v_x,v_y)$, is the "optical transfer function" of the system. The Fourier transform thus reduces an integration to a much simpler multiplication.

If the noise component is negligible, then inversion of Eq. (8.12) can proceed directly to yield the following estimate for $I(v_x,v_y)$:

$$I^*(v_x,v_y) = O(v_x,v_y)\left[\frac{1}{N\,P(v_x,v_y)}\right] \tag{8.13}$$

The image restoration sequence thus involves Fourier transforming the output, applying the "inverse filter", $[N\, P(v_x,v_y)]^{-1}$, and then inverse Fourier transforming to obtain the estimate $I^*(x,y)$. However, for substantial noise components, application of the inverse filter yields the estimate

$$I^*(v_x,v_y) = I(v_x,v_y) + \left[\frac{n\,(v_x,v_y)}{N\,P(v_x,v_y)}\right] \tag{8.14}$$

At higher frequencies, where $n(v_x,v_y)$ is large and $P(v_x,v_y)$ is small, a substantial high frequency noise component will degrade the resulting estimate, $I^*(x,y)$. For these cases, a more sophisticated method of deconvolution is required (B.4, B.11).

8.3 MODULATION TRANSFER FUNCTIONS

A great body of knowledge exists on the general problem of image restoration. Our intent here is not to pursue this line of analysis but rather to describe some fundamental frequency domain descriptions relevant to imaging systems, in particular, to neutron radiography. An important and widely utilized description is the "modulation transfer function", defined as the modulus of the complex optical transfer function (B.16, B.17). Mathematically, we have

$$\mathrm{MTF}\,(v_x,v_y) = |P(v_x,v_y)| \tag{8.15}$$

For an isotropic system, Sec. 6.2, $P(v_x,v_y)$ is real (B.17) and thus becomes $\mathrm{MTF}(v_x,v_y)$. In this case, MTF( ) is more directly calculable as the zeroth order Hankel transform of $P(x,y,u,v)$, given by

$$\mathrm{MTF}\,(v_R) = 2\pi \int_0^\infty P(R)\,J_0\,(2\pi\, v_R\, R)\,R\,dR \tag{8.16}$$

where J_0 is the zeroth order Bessel function (B.6) and

$$R^2 = (x-u)^2 + (y-v)^2 \tag{8.17a}$$

$$v_R{}^2 = v_x{}^2 + v_y{}^2 \tag{8.17b}$$

Physically, $\mathrm{MTF}(v_x,v_y)$ is interpreted as the ratio of the output amplitude to input amplitude of a sinusoidal pattern with rectangular spatial frequencies v_x and v_y being passed through the imaging system.

For one-dimensional imaging, the relevant imaging equation is

$$O(x) = N\,I(x) * L(x) + n(x) \tag{8.18}$$

where $L(x)$ is the shift invariant line-spread function whose Fourier transform is the one-dimensional modulation transfer function, $\mathrm{MTF}(v_x)$. Modulation transfer functions corresponding to line and point-spread functions used variously throughout this text are summarized in Table 8.1, and displayed in Fig. 8.1. Several characteristics of these functions merit discussion. First, MTF( ) generally decreases with spatial frequency, consistent with blurring effects as well as the smoothing of radiographic noise due to information spreading and masking incurred by image degradation sources. Second, a MTF( ) is always less than unity for these image degradation sources. At zero frequency, the scattering MTF( ) is smaller than unity due to backscattering losses. Also, since the uncollided beam component is, by definition, undegraded by scattering, this MTF( ) possesses a base component at all frequencies. Finally, an ideal MTF( ), corresponding to an unrealizable imaging system with no image degradation, has an MTF( ) of unity at all frequencies.

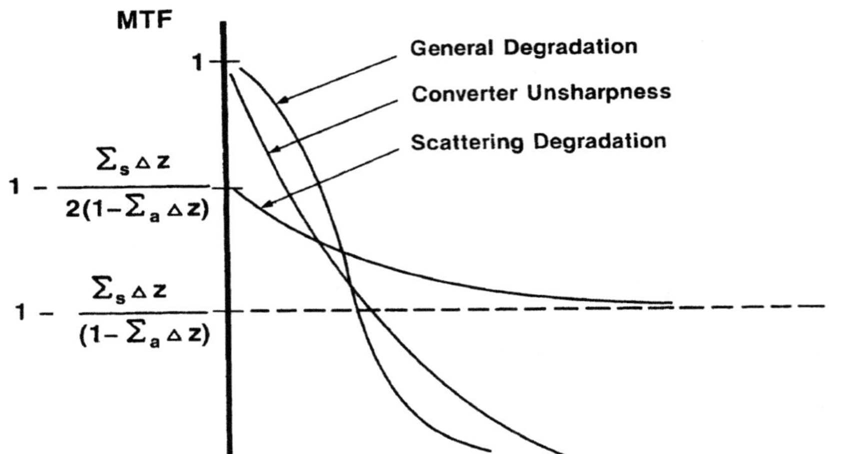

Fig. 8.1: Modulation transfer functions corresponding to the image degradation sources of Table 8.1.

8.4 COMBINED IMAGE DEGRADATION

Given the tools of linear systems analysis, it is possible to compactly describe the combined effects of several image degradation sources (A.61). Without reference to specific degradation sources, we will denote the shift invariant point-spread function representing the

Image Degradation Source	Line or Point-Spread Function (L or P)	Modulation Transfer Function
Converter Unsharpness (Sec. 3.3)	$L = \frac{\lambda}{\pi}[1 + \lambda^2(x-u)^2]^{-1}$	$\exp[-2\pi\lvert v_x\rvert/\lambda]$
Converter Unsharpness (Sec. 6.2)	$P = \frac{\lambda^2}{2\pi}[1 + \lambda^2 R^2]^{-3/2}$	$\exp[-2\pi\lvert v_R\rvert/\lambda]$
Scattering Degradation (Sec. 7.3)	$L_s = [1-\Sigma_a\Delta z]^{-1}\left\{[1-\Sigma_t\Delta z]\,\delta(v-v_o) + \frac{\Sigma_s\Delta z}{2\pi d_o}\left[1+\left(\frac{v-v_o}{d_o}\right)^2\right]^{-1}\right\}$	$[1-\Sigma_a\Delta z]^{-1}\left\{[1-\Sigma_t\Delta z] + \left(\frac{\Sigma_s\Delta z}{2}\right)\exp[-2\pi\lvert v_v\rvert d_o]\right\}$
Scattering Degradation (Sec. 7.3)	$P = [1-\Sigma_a\Delta z]^{-1}\left\{[1-\Sigma_t\Delta z]\,\delta(R) + \frac{\Sigma_s\Delta z}{4\pi d_o^2}\left[1+\left(\frac{R}{d_o}\right)^2\right]^{-3/2}\right\}$	$[1-\Sigma_a\Delta z]^{-1}\left\{[1-\Sigma_t\Delta z] + \left(\frac{\Sigma_s\Delta z}{2}\right)\exp[-2\pi\lvert v_R\rvert d_o]\right\}$
General Degradation (Sec. 3.3)	$L_{gen} = \frac{\alpha}{\sqrt{\pi}}\exp[-\alpha^2 x^2]$	$\frac{\alpha}{\sqrt{\pi}}\exp[-\alpha^2 v_x^2]$

Table 8.1: Modulation transfer functions corresponding to line and point-spread functions used throughout this text.

j^{th} of M image degradation sources by $P_j(\ )$. Then an original undegraded object plane neutron intensity distribution, I( ), is serially transformed through the M image degradation processes to become the final output, O_M, as indicated in Fig. 8.2a.

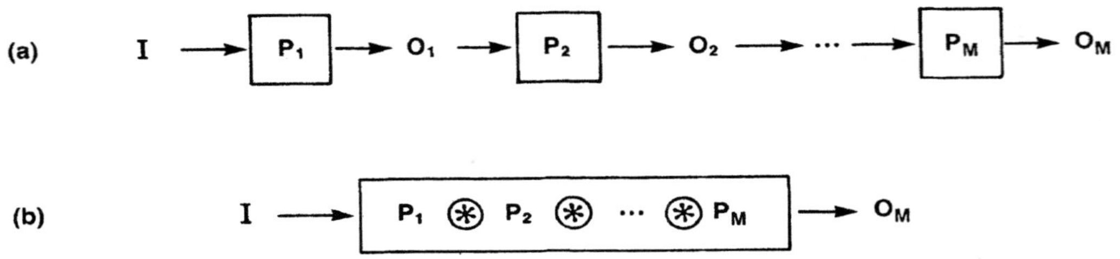

Fig. 8.2: Depiction of image degradation formulations: (a) serial degradation with point-spread function, P_j, transforming an object plane input, I, into an image plane output, O_M; (b) equivalent point-spread function is the convolution of individual point-spread functions.

The final output may be the secondary radiation exposure on the film emulsion or perhaps the optical density variation. This sequence of transformations can be represented mathematically by the following set of coupled linear convolution equations, in accordance with Eq. (8.11), for the noise-free component of the image:

$$\begin{aligned} O_1 &= N_1 I * P_1 \\ O_2 &= N_2 O_1 * P_2 \\ &\ \vdots \\ O_M &= N_M O_{M-1} * P_M \end{aligned} \tag{8.19}$$

Here, O_j is the intermediate output from the input O_{j-1} being transformed by the jth image degradation source and N_j is an arbitrary normalization. Combining Eqs. (8.19) leads to

$$O_M = N_M (N_{M-1} \dots (N_1 I * P_1) * \dots * P_{M-1}) P_M \tag{8.20}$$

which, by associativity of the convolution operation, can be written as

$$O_M = N\, I * P \tag{8.21}$$

where the normalization constant, N, and overall point-spread function, P( ), are defined as

$$N = N_1\, N_2 \ldots N_M \tag{8.22}$$

$$P = P_1 * P_2 * \ldots * P_M \tag{8.23}$$

Thus, the overall system point-spread function representing a combination of image degradation sources, each of which is represented by a separate point-spread function, is simply the convolution of the individual point-spread functions. The equivalence of Eqs. (8.19) and (8.21) is depicted in Fig. 8.2.

For one-dimensional imaging, the overall system line-spread function is the convolution of line-spread functions representing the individual image degradation sources. One interesting implication of this result arises when each of the component line-spread functions is modelled using either of the two following functional forms, Sec. 3.3:

$$L_{\lambda j}(x) = \frac{\lambda_j}{\pi}\,[1 + \lambda_j^2\, x^2]^{-1} \tag{8.24a}$$

$$L_{\alpha j}(x) = \frac{\alpha_j}{\sqrt{\pi}}\, \exp\,[-\,\alpha_j^2\, x^2] \tag{8.24b}$$

In Eqs. (8.24), the subscripts λ and α refer to the functional form of L( ) while the subscript j refers to the jth image degradation source. For these two functional forms, the overall line-spread function is

$$L_\lambda(x) = L_{\lambda 1}(x) * L_{\lambda 2}(x) * \ldots * L_{\lambda M}(x) \tag{8.25a}$$

$$L_\alpha(x) = L_{\alpha 1}(x) * L_{\alpha 2}(x) * \ldots * L_{\alpha M}(x) \tag{8.25b}$$

Fourier transformation of Eqs. (8.25) using the convolution theorem, Sec. 8.2, results in

$$MTF_\lambda(v_x) = \prod_{j=1}^{M} MTF_{\lambda j}(v_x) \tag{8.26a}$$

$$MTF_\alpha(v_x) = \prod_{j=1}^{M} MTF_{\alpha j}(v_x) \tag{8.26b}$$

where $MTF_{\lambda j}(\ )$ and $MTF_{\alpha j}(\ )$ are the modulation transfer functions derived by Fourier transforming $L_{\lambda j}(\ )$ and $L_{\alpha j}(\ )$. Substituting for the $MTF_{\lambda j}(\ )$ and $MTF_{\alpha j}(\ )$ from Table 8.1 yields

$$MTF_{\lambda}(v_x) = \exp\left[-2\pi|v_x| \prod_{j=1}^{M} \frac{1}{\lambda_j}\right] \tag{8.27b}$$

$$MTF_{\alpha}(v_x) = \frac{\alpha}{\sqrt{\pi}} \exp\left[-v_x^2 \prod_{j=1}^{M} \alpha_j^2\right] \tag{8.27b}$$

Inverse Fourier transformation of Eqs. (8.27) yields, upon inspection,

$$L_{\lambda}(x) = \frac{\lambda}{\pi}[1 + \lambda^2 x^2]^{-1} \tag{8.28a}$$

$$L_{\alpha}(x) = \frac{\alpha}{\sqrt{\pi}} \exp[-\alpha^2 x^2] \tag{8.28b}$$

where the overall line-spread parameters are defined by

$$\frac{1}{\lambda} = \sum_{j=1}^{M} \frac{1}{\lambda_j} \tag{8.29a}$$

$$\alpha^2 = \sum_{j=1}^{M} \alpha_j^2 \tag{8.29b}$$

This modelling framework provides a useful additive property for image degradation analysis. Specifically, image degradation is additive in the sense that each of the M degradation parameters, λ_j or α_j, sum according to Eqs. (8.29) to give the overall parameter, λ or α. This addition property also holds for the corresponding isotropic point-spread functions, $P_{\lambda j}(R)$ and $P_{\alpha j}(R)$.

8.5 SCATTERING DEGRADATION AND CONVERTER UNSHARPNESS

The combination of image degradation sources represented by a convolution of transfer functions can be demonstrated for two degradation sources analyzed in some detail in this text, namely, scattering degradation and converter unsharpness. For one-dimensional imaging, the relevant line-spread functions are found in Table 8.1 and are reproduced below using consistent variables, with $L_c(\ )$ representing converter unsharpness and $L_s(\ )$ representing single interaction object scattering.

$$L_c(x, u) = \frac{\lambda}{\pi}[1 + \lambda^2(x - u)^2]^{-1} \tag{8.30}$$

$$L_s(u, u_o) = [1 - \Sigma_a \Delta z]^{-1}\left\{[1 - \Sigma_a \Delta z]\delta(u - u_o) + \frac{\Sigma_s \Delta z}{2\pi d_o}\left[1 + \left(\frac{u - u_o}{d_o}\right)^2\right]^{-1}\right\} \tag{8.31}$$

The information flow described by Eqs. (8.30) and (8.31) is shown schematically in Fig. 8.3. An uncollided line source of neutrons with unit intensity exits the object at the line u_o in the object plane. Object scattering degrades this ideal "input" so that a neutron intensity of $L_s(u, u_o)$ results at the line u in the object plane. Taking the object plane to be as close as possible to the converter, the neutron intensity at the line u will spread through converter unsharpness to yield a certain secondary radiation exposure at the line x in the film emulsion. The relative magnitude of this exposure is given by the combined line-spread function

$$\begin{aligned} L(x, u_o) &= L_c * L_s \\ &= \int_{-\infty}^{\infty} L_c(x - u)L_s(u - u_o)du \end{aligned} \tag{8.32}$$

By the convolution theorem, the MTF( ) corresponding to $L(x, u_o)$ is the product of the MTF( )'s for $L_c(x-u)$ and $L_s(u-u_o)$. Substituting these MTF( )'s from Table 8.1 and taking the inverse Fourier transform of their product results in

$$L(x, u_o) = [1 - \Sigma_a \Delta z]^{-1}\left\{[1 - \Sigma_a \Delta z]\, L_c(x, u_o) + \left(\frac{\Sigma_s \Delta z}{2}\right)\left(\frac{1}{\pi d}\right)\left[1 + \left(\frac{x - u_o}{d}\right)^2\right]^{-1}\right\} \tag{8.33}$$

where the combined degradation parameter is defined as the following sum of the individual degradation parameters:

$$d = d_o + \frac{1}{\lambda} \tag{8.34}$$

The combined line-spread function, L( ), will always be broader than either of its components, $L_c(\ )$ and $L_s(\ )$.

Consider next a specific application of the combined line-spread function. In similarity to Sec. 7.5, the image plane converter response resulting from a normally incident collimated neutron beam of uniform intensity, I_o, traversing a knife-edge slab object, with single scattering and converter unsharpness image degradation, is the following convolution:

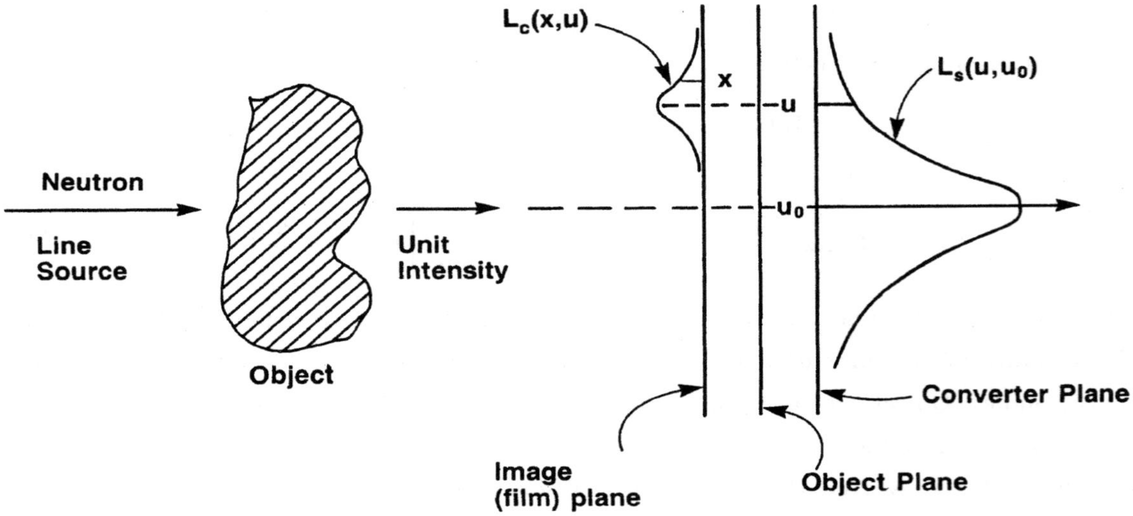

Fig. 8.3: Line source of neutrons at u_0 in the object plane undergoing scattering degradation, $L_s(u,u_0)$, and converter unsharpness, $L_c(x,u)$.

$$E(x) = N \int_{-\infty}^{\infty} I(u_0)\, L^*(x,u_0)\, du_0 \tag{8.35}$$

where L*( ) is the overall line-spread function. For ease of reference, the slab geometry is reproduced in Fig. 8.4. In Eq. (8.35), N is an arbitrary normalization constant and I( ) is the input function

$$I(u_0) = \begin{cases} I_0(1-\Sigma_a \Delta z), & u_0 < 0 \\ I_0, & u_0 > 0 \end{cases} \tag{8.36}$$

Also, there are two image degradation processes for $u_0 < 0$, the slab region, but only converter unsharpness for $u_0 > 0$. Consequently the relevant overall line-spread function is

$$L^*(x,u_0) = \begin{cases} L(x,u_0), & u_0 < 0 \\ L_c(x,u_0), & u_0 > 0 \end{cases} \tag{8.37}$$

An evaluation of Eq. (8.35) yields an edge-spread function such as that given in Fig. 8.5 (A.62). A comparison of Fig. 8.5 and Fig. 7.6 reveals that object scattering induces the hump-shaped edge response distortions while converter unsharpness diminishes them through smoothing.

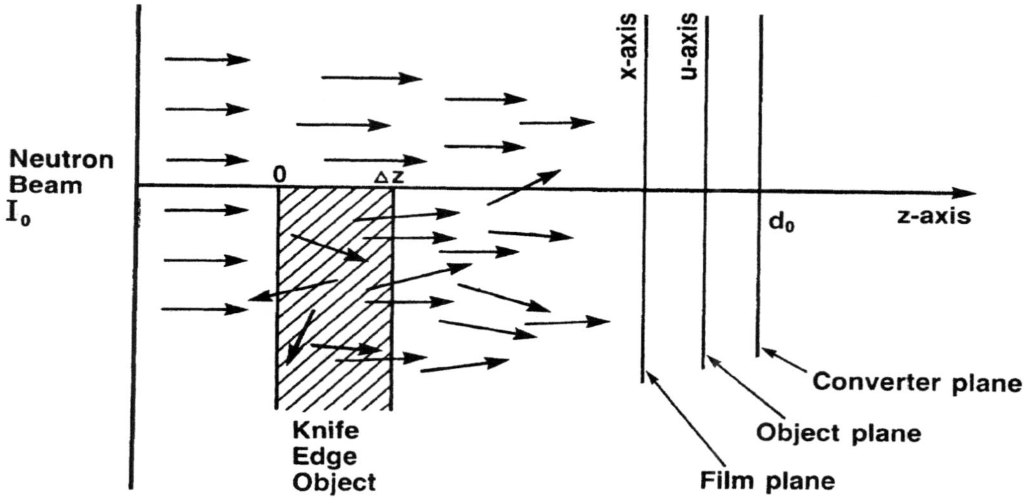

Fig. 8.4: Neutron radiographic processes for a knife-edge object incorporating scattering degradation and converter unsharpness.

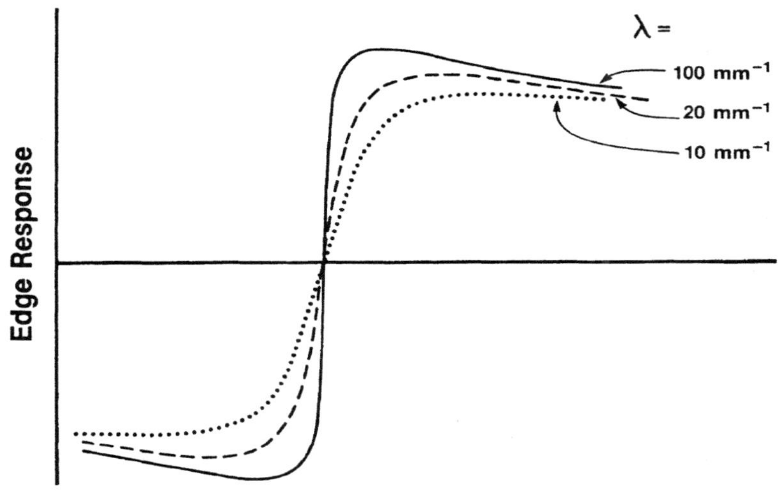

Fig. 8.5: Edge-spread function of a knife-edge object in which neutron absorption and scattering combines with converter unsharpness to yield an edge accentuation effect.

IX. Selected Topics

The topics of the preceding chapters were chosen in order to allow a coherent sequential development of our subject. The following additional topics provide further physical insight into, and mathematical descriptions of, neutron radiography.

9.1 NEUTRON-NUCLEUS INTERACTIONS

In Secs. 2,3 and 2.4, the neutron-nucleus interaction cross sections were introduced as interaction proportionality constants characterizing the rate of absorption (σ_a, Σ_a) and scattering (σ_s, Σ_s) in the medium of interest. In reality, the interaction probabilities depend not only on the isotopes constituting the medium, but also on the energy of the neutrons interacting with the medium. Furthermore, several types of neutron scattering and absorption are possible.

A neutron colliding with a nucleus may, for example, scatter elastically, imparting energy and momentum to the nucleus and then leaving in a different direction with an altered energy, or scatter inelastically, leaving the nucleus in an excited state. If the neutron is absorbed, the nucleus may de-excite by emitting a γ-ray, a charged particle or even multiple neutrons, although this latter process is quite rare and occurs only at very high neutron energies. The cross sections, $\sigma_{a,i}$ and $\sigma_{s,i}$, in Eqs. (2.15) are thus the sums of individual microscopic cross sections characterizing the various scattering and absorption processes.

The energy dependence of cross sections is usually incorporated by integration, in such a way that for each energy group of neutrons, a single energy-averaged cross section is established (B.10). For radiography with thermal neutrons, a sufficient description can be achieved by employing only a single energy group containing all relevant energies, as defined by the following "one-group" cross sections:

$$\sigma_{a,i} = \frac{\int_0^\infty \sigma_{a,i}(E)\, \psi(E)\, dE}{\int_0^\infty \psi(E)\, dE} \tag{9.1a}$$

$$\sigma_{s,i} = \frac{\int_0^\infty \sigma_{s,i}(E)\, \psi(E)\, dE}{\int_0^\infty \psi(E)\, dE} \tag{9.1b}$$

$$\Sigma_{a,i}(z) = \frac{\int_0^\infty \Sigma_{a,i}(E, z)\, \psi(E)\, dE}{\int_0^\infty \psi(E)\, dE} \tag{9.1c}$$

$$\Sigma_{s,i}(z) = \frac{\int_0^\infty \Sigma_{s,i}(E, z)\, \psi(E)\, dE}{\int_0^\infty \psi(E)\, dE} \tag{9.1d}$$

Here, $\psi(E)$ is the energy spectrum of the neutrons interacting with the medium. For example, in a large homogeneous medium in which scattering dominates absorption and the source neutrons are in near thermal equilibrium with their source environment, $\psi(E)$ for the incident beam can be approximated by the Maxwellian distribution function

$$\psi(E) \sim E \exp[-E/kT] \tag{9.2}$$

where k is the Boltzmann constant, and T is the absolute temperature of the medium. However, the neutron spectrum may change with depth of penetration into the medium due to the energy dependence of the cross sections. In general, the accuracy of the one-group description depends on the accuracy with which $\psi(E)$ can be determined.

Note also that scattering cross sections generally possess an angular dependence. In practice, a low-energy neutron-nucleus collision is isotropic in the center-of-mass system and nearly isotropic in the laboratory system, except for low mass scatterers like hydrogen.

9.2 COMPARISON OF LINE-SPREAD FUNCTIONS

The following two line-spread functions are of relevance to neutron radiography (e.g. Sec. 3.3):

$$L_1(x) = \frac{\lambda/\pi}{1 + \lambda^2 x^2} \tag{9.3a}$$

$$L_2(x) = \frac{\alpha}{\sqrt{\pi}} \exp[-\alpha^2 x^2] \tag{9.3b}$$

In particular, $L_1(x)$ arises specifically from converter unsharpness and $L_2(x)$ is a general line-spread function often called a "Gaussian". Other functional forms are of course possible, but the functions in Eqs. (9.3) are attractively simple in form while providing good descriptions. For example, the exponential line-spread function given by

$$L_3(x) = \frac{\gamma}{2} \exp[-\gamma|x|] \tag{9.3c}$$

is simple in form but unrealistic due to its cusp at $x=0$.

A meaningful geometrical comparison between $L_1(x)$ and $L_2(x)$ can be established by considering their shapes when they have equal widths – that is, equal full-widths-at-half-maximum (FWHM). As derived in Sec. 3.4, the FWHM for $L_1(x)$ is

$$\text{FWHM}_1 = 2/\lambda \tag{9.4a}$$

while the FWHM for $L_2(x)$ is similarly found to be

$$\text{FWHM}_2 \simeq 1.665/\alpha \tag{9.4b}$$

From Fig. 9.1, it is evident that $L_2(x)$ is intrinsically narrower than $L_1(x)$. Of course this does not imply that every Gaussian is narrower than every function of the form in Eq. (9.3a), since absolute width depends on the resolution parameters λ and α.

Since $L_2(x)$ is intrinsically narrower than $L_1(x)$, it also follows that the edge-spread function derived from $L_2(x)$, denoted by $S_2(x)$, is intrinsically sharper than that derived from $L_1(x)$, denoted by $S_1(x)$. From Sec. 3.5, integration of $L_1(x)$ leads to

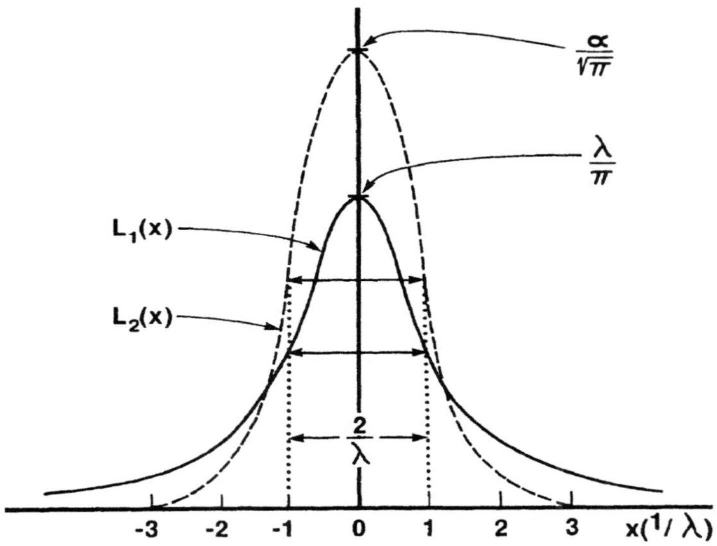

Fig. 9.1: Comparison of different line-spread functions with the same full-width-at-half maximum.

$$S_1(x) = N\left[\frac{1}{2} + \frac{1}{\pi}\tan^{-1}(\lambda x)\right] \tag{9.5a}$$

It can be similarly derived that

$$S_2(x) = N\left[\frac{1}{2} + \frac{1}{2}\,\mathrm{erf}(\alpha x)\right] \tag{9.5b}$$

where erf( ) is the error function (B.6). These two edge-spread functions are depicted in Fig. 9.2 for the case in which the line-spread functions from which they are derived have equal FWHM. Note that while $S_2(x)$ is intrinsically sharper than $S_1(x)$, either functional form can be used to approximate the other if the resolution parameter is appropriately selected. In particular, $S_1(x)$ and $S_2(x)$ are very similar if their resolution parameters are related by

$$\lambda = \left\{\left(\frac{1}{1.255}\right)\tan\left[\frac{\pi}{2}\,\mathrm{erf}(1.255)\right]\right\}\alpha \tag{9.6}$$

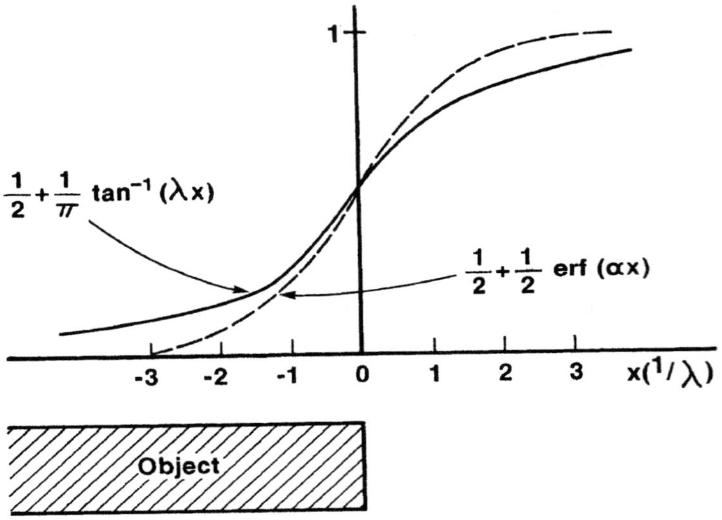

Fig. 9.2: Normalized edge-spread functions derived from the line-spread functions of Fig. 9.1

9.3 IMAGE DEGRADATION CLASSIFICATION

The five major sources of image degradation in neutron radiography, first introduced in Sec. 3.2, are converter unsharpness, scattering degradation, geometric unsharpness, motion unsharpness and noise degradation. In previous chapters, only converter unsharpness and scattering degradation have been analyzed in detail. Before proceeding to discuss in detail the remaining three image degradation sources, it is appropriate to establish a perspective from which the natures of these degradation sources, as well as their mathematical formulations, can be compared.

First, the five sources can be grouped according to the manner in which they degrade image resolution. The "unsharpness" label applies to converter, geometric and motion degradation sources because they decrease resolution by blurring image boundaries. Single parameter characterizations of each unsharpness source have led to conventional formulations for the total, or combined, unsharpness in radiography (A.48, B.18). Object scattering and image noise, however, are not unsharpness sources, but rather have the common property of decreasing resolution by masking the undegraded image. Aside from scattering induced local edge enhancement effects, Sec. 7.5, scattered radiation can indeed be considered as a source of image noise that is always positive. Thus, a uniform background scatter component behaves like noise with a Dirac delta probability density function for the magnitude.

Second, the five sources may be grouped by their mathematical formulations, in particular, by the facility with which each source can be cast as a system transfer function for linear systems analysis. We have seen that both converter unsharpness and simple cases of scattering degradation can be characterized by shift invariant line and point-spread functions. In general, object scattering can be described by a shift variant point-spread function with direct dependence on the ideal image itself and is therefore a nonlinear degradation sytem. In Sec. 9.4, we will see that geometric edge unsharpness is also characterized by a shift variant transfer function. Image noise, however, is probabilistically described and is treated as a component added to the image signal, as opposed to a degradation system with which the ideal image signal is convolved. Finally, in Sec. 9.5, we will see that motion-induced edge unsharpness can be analyzed as a linear degradation system with a shift variant line-spread function, and is therefore mathematically similar to geometric unsharpness.

9.4 GEOMETRIC UNSHARPNESS

At a fundamental level of description, geometric unsharpness arises because the incident beam geometry is such that a given point in an object projects to a region of points in the image, as illustrated in Fig. 9.3. Image regions for adjacent object points thus overlap, resulting in a blurred image.

It is obvious from Fig. 9.3 that the single object point image region is generally proportional to both the source size and the object-to-image plane separation, and inversely proportional to the object-to-collimator plane separation. For well collimated beams, these direct variations are only approximate. It is true, however, that geometric unsharpness can

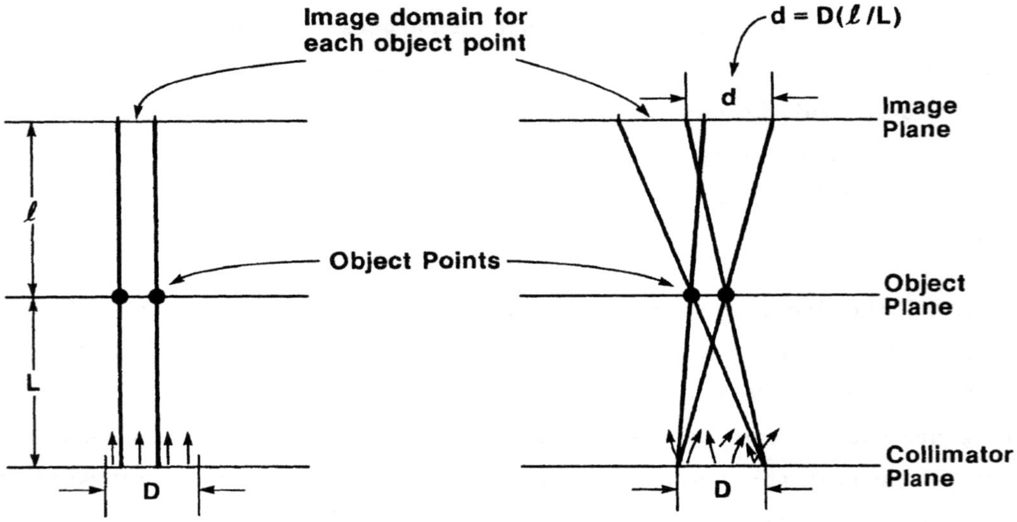

Fig. 9.3: Illustration of geometric unsharpness arising from a beam with finite divergence.

be removed by either effectively placing the collimator plane at infinity or by eliminating the object-to-image plane separation.

The specific beam characteristic responsible for geometric unsharpness is evidently the combination of a finite source area in the collimator plane and a finite beam divergence. Indeed, in the limit as either the source area or beam divergence individually approaches zero, geometric unsharpness is removed. Since neutron radiography is usually performed with a broad beam, it is therefore the beam divergence which must be restricted in order to restrict geometric unsharpness. As discussed in Sec. 2.2, beam divergence is determined by the selected collimator geometry, so that for a given object placement, it is ultimately the collimator that determines the level of geometric unsharpness in obtained images.

The manner in which geometric unsharpness generates edge unsharpness is depicted in Fig. 9.4. Divergent rays from a partially collimated neutron source impinge on a neutronically black object that forms an edge to the beam. The finite beam divergence allows some neutrons to strike the converter directly behind the object, resulting in a blurred edge image. The object edge is defined as being at u = 0 with the incident beam source between a_- and a_+. In practice, the beam collimator restricts the angular variation of the source neutrons to a forward cone of directions.

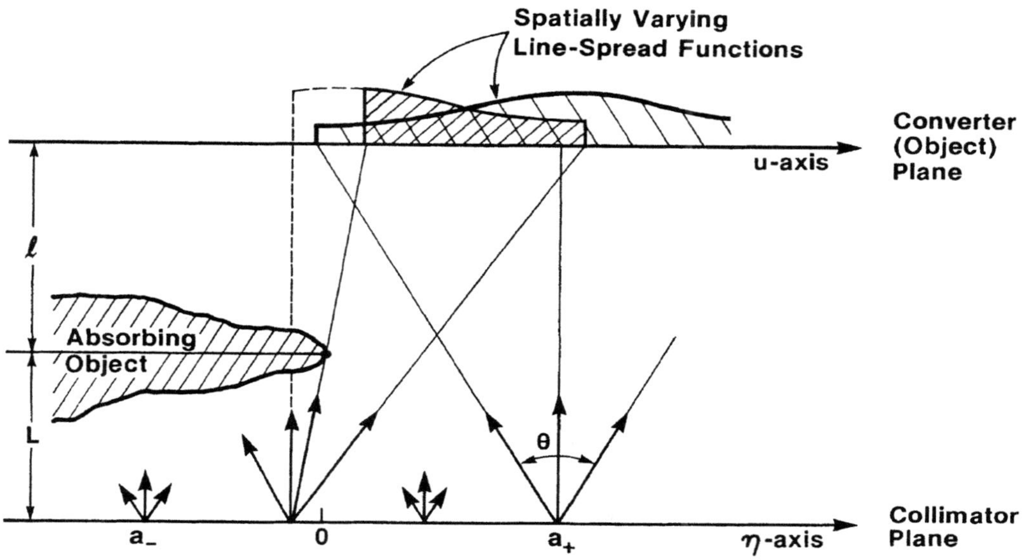

Fig. 9.4: Divergent neutron beam incident on an absorbing object and leading to edge unsharpness.

A line source of neutrons from the general collimator coordinator, η, will thus be distributed according to some line-spread function, $L_1(u,\eta)$, at the general object plane coordinate, u. The object plane is conveniently selected as being adjacent to the converter plane.

The FWHM of $L_1(u,\eta)$ is determined by the neutron source angular distribution and also by the separation, $\ell + L$, of the collimator and converter planes. As suggested in Fig. 9.4, $L_1(u,\eta)$ is shift variant, consisting of a section of general line-spread function, $L_1{}^*(u,\eta)$, that would be obtained in the absence of the absorbing object. If the source neutrons are angularly distributed isotropically within the forward direction cone, then $L_1{}^*(u,\eta)$ is given, Sec. 3.3, by

$$L_1^*(u,\eta) = \begin{cases} \dfrac{\lambda_1/\pi}{1+\lambda_1^2(u-\eta)^2}, & |u-\eta| \le \dfrac{1}{\lambda_1}\tan\left(\dfrac{\theta}{2}\right) \\ 0, & |u-\eta| > \dfrac{1}{\lambda_1}\tan\left(\dfrac{\theta}{2}\right) \end{cases} \tag{9.7}$$

where the spreading parameter, λ_1, is the inverse separation given by

$$\lambda_1 = \frac{1}{\ell + L} \tag{9.8}$$

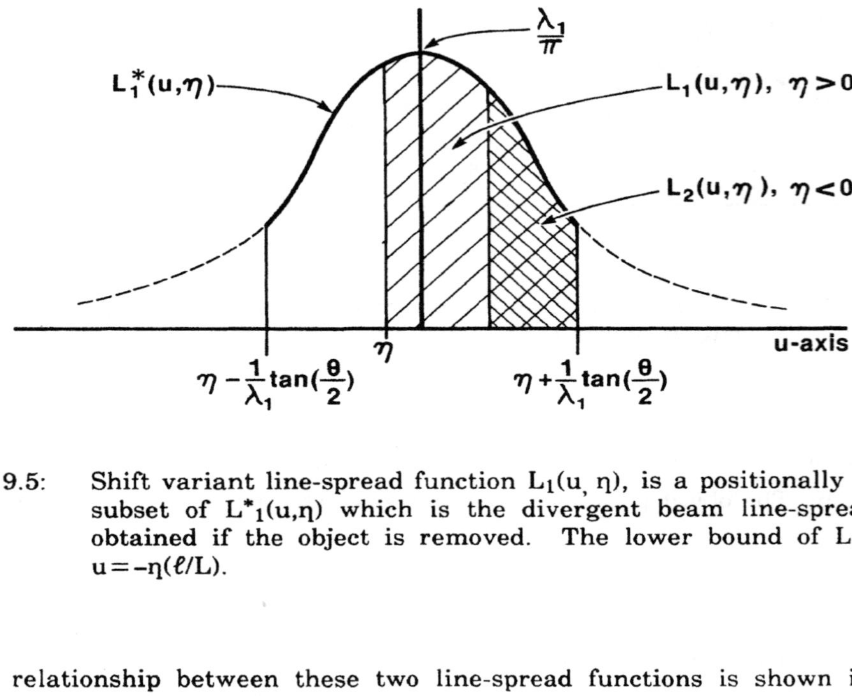

Fig. 9.5: Shift variant line-spread function $L_1(u,\eta)$, is a positionally determined subset of $L^*_1(u,\eta)$ which is the divergent beam line-spread function obtained if the object is removed. The lower bound of $L_1(u,\eta)$ is at $u=-\eta(\ell/L)$.

The relationship between these two line-spread functions is shown in Fig. 9.5. Referring back to Fig. 9.4, the object plane intensity, $\phi_t(u)$, is the sum of contributions from each of the line sources comprising the collimator plane source. Mathematically, $\phi_t(u)$ is the following convolution:

$$\phi_t(u) = \int_{-\infty}^{\infty} I(\eta)\, L_1(u,\eta)\, d\eta \tag{9.9}$$

Here, $L_1(\ )$ and $L_1^*(\ )$ are related, Fig. 9.5, by

$$L_1(u,\eta) = \begin{cases} L_1^*(u,\eta)\,, & \eta \in \left[-u\left(\frac{L}{\ell}\right), a_+\right] \\ 0\,, & \text{otherwise} \end{cases} \tag{9.10}$$

and the collimator plane intensity distribution, $I(\eta)$, is given by

$$I(\eta) = \begin{cases} I_o\,, & \eta \in [a_-, a_+] \\ 0\,, & \text{otherwise} \end{cases} \tag{9.11}$$

Evaluation of Eq. (9.9) yields an edge response such as that shown in Fig. 9.6. Unlike converter unsharpness, the edge response for geometric unsharpness vanishes at a fixed distance, d*, from the actual edge location, behind the absorbing object. This distance, and hence the magnitude of the penumbral region, increases with beam divergence, θ, and the object-to-converter separation, ℓ, as follows

$$d^* = \ell \tan\left(\frac{\theta}{2}\right) \tag{9.12}$$

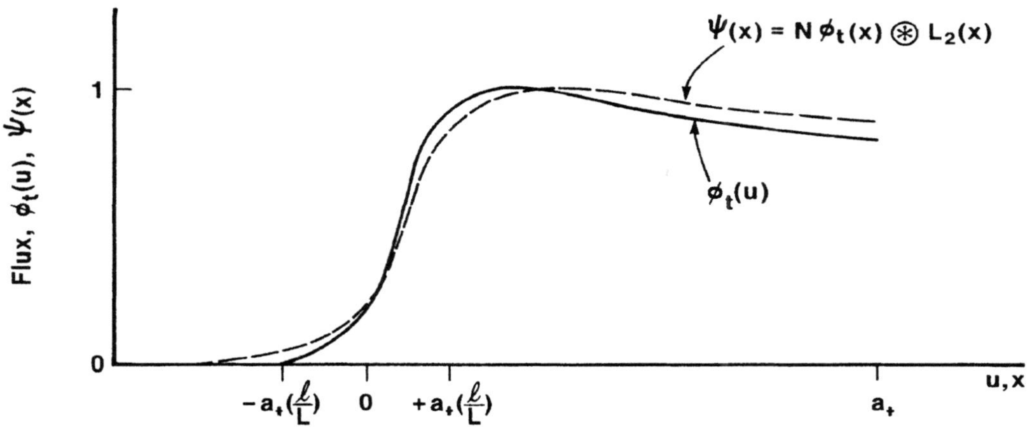

Fig. 9.6: Geometric unsharpness resulting in object plane (–) and converter response (- - -) penumbra for a totally absorbing object.

Note that geometric unsharpness will be increased if the object is only partially absorbing, and is compounded by converter unsharpness. If $L_2(x,u)$ is the converter unsharpness line-spread function, then the film blackening secondary radiation exposure combining geometric and converter unsharpness is given by the following convolution:

$$\psi(x) = N \int_{-\infty}^{\infty} I(\eta)\, L(x, \eta)\, d\eta \tag{9.13}$$

Here, N is an arbitrary normalization constant and L(x) is the overall shift variant line-spread function, given by

$$L(x, \eta) = \int_{-\infty}^{\infty} L_1(u, \eta)\, L_2(x, u)\, du \tag{9.14}$$

9.5 SINGLE MOTION UNSHARPNESS

Motion unsharpness can arise when any of the three fundamental imaging chain components – the source, object and recorder – move relative to one or both of the others during the exposure period. In conventional neutron radiography, motion unsharpness normally results from object motion alone. Since even thermal neutrons travel at high speeds ($\approx$ 2200 m/s) relative to the mechanical motion speeds normally encountered with life-sized objects, it is usually an excellent approximation that motion unsharpness is solely a function of the object motion, possessing no dependence on the neutron transit time between the collimator and converter plane. We therefore employ this assumption in order to analyze the single motion model shown in Fig. 9.7.

In this model, an absorbing object edge is positioned at the coordinate x_0 for t_0 seconds, and then moves with speed v(t) in the negative x-direction to arrive, at time t_1, at the following x - coordinate:

$$x_1 = x_0 + \int_{t_0}^{t_1} v(t)\, dt \tag{9.15}$$

The edge then remains positioned at x_1 until time τ, when the exposure is terminated. This model is therefore relevant to the radiographic situation in which a single, undirectional object motion occurs during the exposure.

Since the object plane neutron intensity distribution, $\phi_t(u,t)$, is time dependent, it is the temporally integrated intensity distribution, also referred to as the fluence, that ultimately generates the optical density variation. Mathematically, the fluence is given by

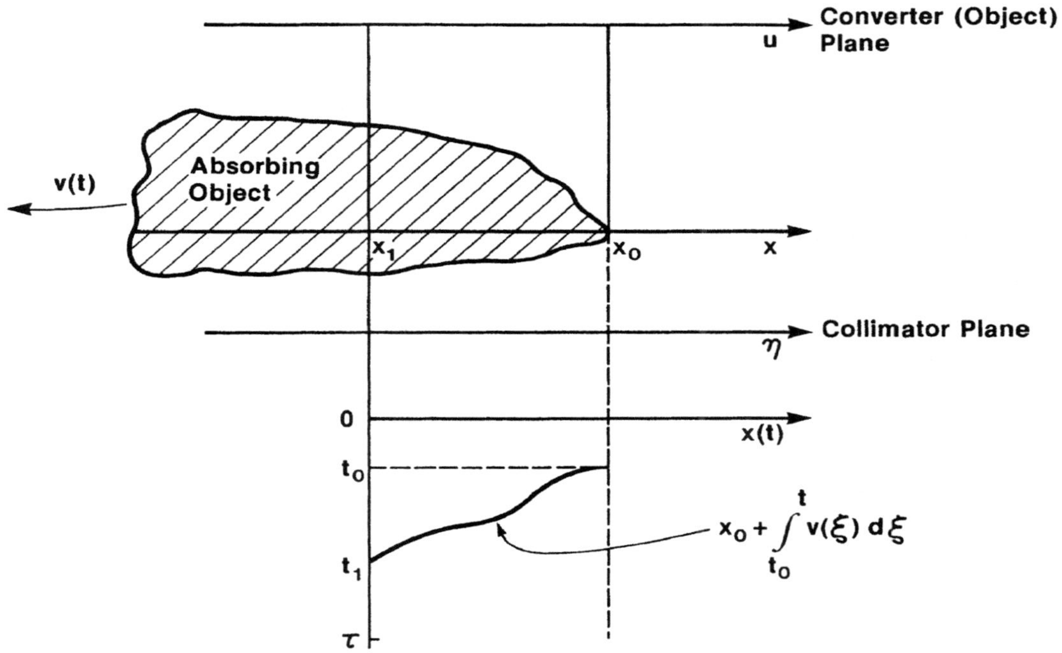

Fig. 9.7: Absorbing edge moving from x_0 to x_1 between time t_0 and t_1 with velocity v(t).

$$\Phi_\tau(u) = \int_0^\tau \Phi_t(u, t)\, dt \tag{9.16}$$

It is convenient to divide the above integration into the three periods governing the edge position as follows:

$$\Phi_\tau(u) = I_0 \left\{ \int_0^{t_0} H(u - x_0) dt + \int_{t_0}^{t_1} H(u - u_{edge}(t))\, dt + \int_{t_1}^{\tau} H(u - x_1) dt \right\} \tag{9.17}$$

where H(u) is the Heaviside step function (B.6), given by

$$H(u) = \begin{cases} 1, & u > 0 \\ 0, & u < 0 \end{cases} \tag{9.18}$$

and $u_{edge}(t)$ denotes the u-coordinate of the object edge at time t, given by

$$u_{edge}(t) = \begin{cases} x_0, & t \in (0, t_0) \\ x_0 + \int_{t_0}^{t} v(\xi)\, d\xi, & t \in (t_0, t_1) \\ x_1, & t \in (t_1, \tau) \end{cases} \tag{9.19}$$

Now the first and third integrals in Eq. (9.17) are trivial to evaluate, since their integrands are constants. The second integral can be transformed into a spatial convolution integration of an input function with a shift variant response function by changing variables from t to $u_{edge}(t)$, related through v(t) by

$$v(t) = \frac{d\, u_{edge}(t)}{dt} \tag{9.20}$$

This allows Eq. (9.17) to be rewritten as

$$\Phi_\tau(u) = I_0 t_0 H(u - x_0) + \int_{-\infty}^{\infty} I_0(u_{edge}) R(u, u_{edge}) du_{edge}$$

$$+ I_0(\tau - t_1) H(u - x_1) \tag{9.21}$$

where the input, $I_0(u_{edge})$, is simply the constant collimator plane source intensity I_0. The response function, $R(u, u_{edge})$, is given by

$$R(u, u_{edge}) = \begin{cases} \dfrac{H(u - u_{edge})}{-v(u_{edge})}, & u_{edge} \in (x_1, x_0) \\ 0, & \text{otherwise} \end{cases} \tag{9.22}$$

Physically, $R(u, u_{edge})$ represents the object plane intensity at u, while $v(u_{edge})$ denotes the object velocity, when the object edge is positioned at u_{edge}. Note that $I_0(u_{edge})$ is always constant and that u_{edge} always lies between x_1 and x_0, so that the second line in Eq. (9.22) is only of formal significance. These formal definitions have been employed to allow the second integral in Eq. (9.17) to be written as a true convolution. For constant velocity motion, $R(u,u_{edge})$ becomes shift invariant.

The object plane intensity can now be written as the convolution of a constant input intensity, I_0, with a shift variant overall response function, $R^*(u,s)$, given by

$$R^*(u, s) = R(u, s) + \delta(u - s)[t_0 H(u - x_0) + (\tau - t_1) H(u - x_1)] \tag{9.23}$$

where $\delta(u-s)$ is the Dirac delta distribution. In compact form, $\Phi_\tau(u)$ is thus given by

$$\phi_\tau(u) = \int_{-\infty}^{\infty} I_0 R^*(u, s)\, ds \tag{9.24}$$

While the above analysis is not the most direct way to calculate $\phi_\tau(u)$, it has the important advantage of casting motion unsharpness into a general linear systems framework. This is of considerable utility if the model is to be expanded to compound motion unsharpness with other degradation sources. For example, the film blackening secondary radiation fluence, $\psi_\tau(x)$, is related to ϕ_τ through convolution

$$\psi_\tau(x) = N\, \phi_\tau(x) * L_1(x) \tag{9.25}$$

where $L_1(x)$ is a converter unsharpness line-spread function and N is an arbitrary normalization constant. Combining Eqs. (9.24) and (9.25) yields

$$\psi_\tau(x) = N I_0 \int_{-\infty}^{\infty} L(x, s)\, ds \tag{9.26}$$

where the combined line-spread function, L(x, s), is given by

$$L(x, s) = \int_{-\infty}^{\infty} R^*(u, s)\, L_1(x - u)\, du \tag{9.27}$$

An example of a secondary radiation fluence, $\psi_\tau(x)$, for constant velocity motion is provided in Fig. 9.8. For illustration purposes, $\psi_\tau(x)$ is divided into its three components corresponding to the three time periods logically comprising the object motion. The amplitude of each component is thus proportional to the duration of its time period. Note that the motion component, ψ_2, will always exhibit the greatest spread.

9.6 PERIODIC MOTION UNSHARPNESS

The motion unsharpness analysis in Sec. 9.5 can be generalized to allow bidirectional object motion with only a partially absorbing object at the expense, however, of mathematical simplicity. In this section, we examine the fluence responses to an absorbing object with an edge oscillating sinusoidally between two limits. The analysis is of potential relevance to the neutron radiographic inspection of an encased or shielded machine component undergoing periodic motion.

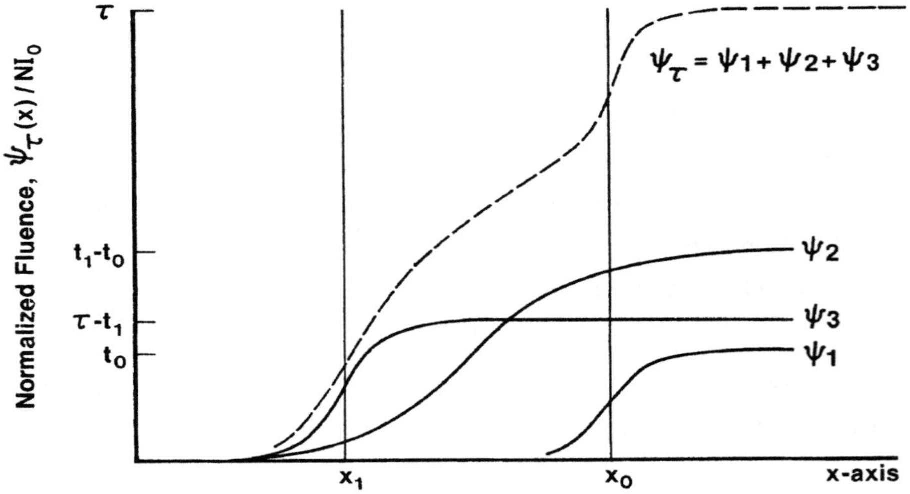

Fig. 9.8: Normalized secondary radiation fluence distribution for an absorbing object moving with constant velocity v from x_0 to x_1 in time (t_1-t_0). The object initially rests at x_0 for all $t<t_0$ and finally rests at x_1 until end of exposure. The above three components are associated with the following time intervals: ψ_1 for $t \in [O, t_0)$, ψ_2 for $t \in [t_0, t_1)$ and ψ_3 for $t \in [t_1, \tau)$.

The model in Fig. 9.7 can be extended to accommodate an oscillating object as follows. First, the object is assumed to be oscillating in steady state, with the edge oscillating between x_0 and x_1, and with no initial or final stationary period. Second, under the assumption that many cycles occur during the exposure period, the fluence distributions can be approximated by those obtained during a single "downstroke", or half-cycle, of the oscillating edge. We can therefore set $t_0 = 0$ and $t_1 = \tau$, with the sinusoidal oscillations described by

$$v(t) = -\frac{\pi}{2\tau}(x_0 - x_1)\sin[\pi(t/\tau)] \tag{9.28a}$$

$$u_{edge}(t) = x_0 - \left(\frac{x_0 - x_1}{2}\right)(1 - \cos[\pi(t/\tau)]) \tag{9.28b}$$

For this case, Eq. (9.17) reduces to

$$\Phi_\tau(u) = I_0 \int_0^\tau H(u - u_{edge}(t))\,dt \tag{9.29}$$

Substitution of Eq. (9.28b) into Eq. (9.29) and evaluation yields, after some algebra

$$\Phi_\tau(u) = I_0 \tau \left\{ 1 - \frac{1}{\pi} \cos^{-1}\left[1 - 2\frac{(x_0 - u)}{(x_0 - x_1)} \right] \right\} \tag{9.30}$$

The object plane fluence is plotted in Fig. 9.9. Also included is a typical secondary radiation fluence, $\psi_\tau(x)$, obtained by convolving $\Phi_\tau(u)$ with a converter unsharpness line-spread function.

Note that if the oscillations occur with a velocity that is more constant than sinusoidal, then the fluence distributions will be smoother near the oscillation boundaries than is indicated in Fig. 9.9. Indeed, a microdensitometer scan parallel to the direction of object motion can potentially be examined to determine the type, regularity and amplitude of the oscillations.

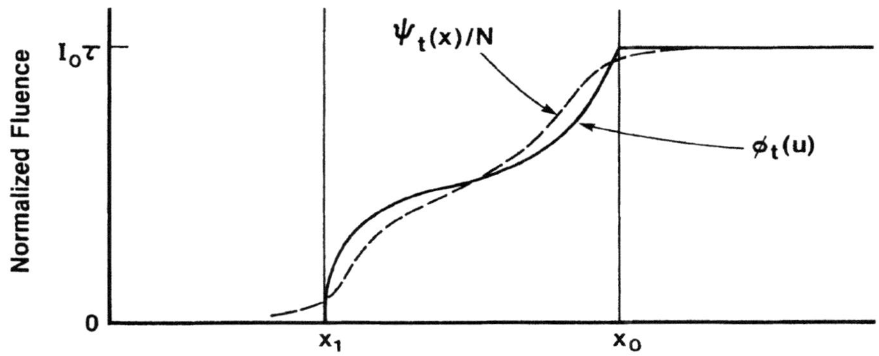

Fig. 9.9: Normalized object plane fluence, $\Phi_t(u)$, and secondary radiation flue $\psi_t(x)$, for an absorbing object with an edge sinusoidally oscilla between x_0 and x_1.

9.7 NOISE DEGRADATION

The addition of noise to either a spatial or temporal signal degrades the signal by limiting the accuracy with which it can be interpreted. This uncertainty extends to the quantitative assessment of output signals, such as the edge response functions discussed in previous chapters. Consider, as an example, the noise-free and noisy edge-spread functions in Fig. 9.10, for which the objective is to determine the edge location through identification of the maximum slope, Sec. 4.2. Without some form of smoothing, the noise component serves to restrict the estimated maximum slope coordinate to a finite range.

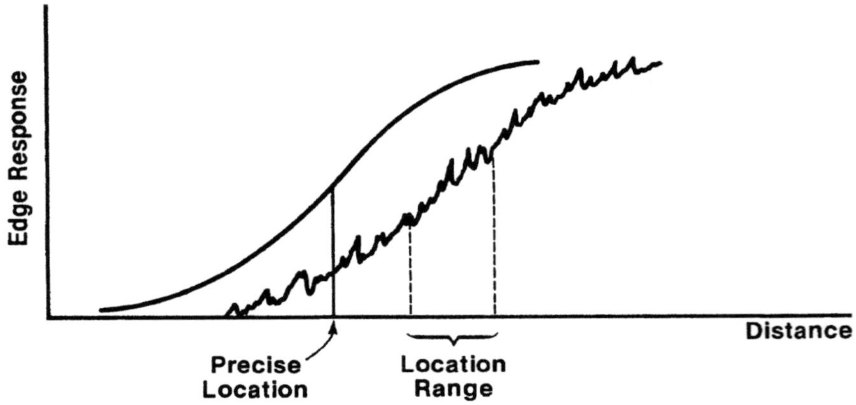

Fig. 9.10: Noise induced uncertainty in determining the edge location coordinate from an edge-spread function.

The problem is further compounded by the procedure of differentiation, since differentiation tends to accentuate the higher spatial frequencies that typically comprise the noise component. It is therefore essential that edge scans be smoothed prior to differentiation, as noted in Sec. 4.2.

There are three important noise sources in neutron radiography, each of which are characteristic of most forms of radiographic imaging. First, there is film-grain noise, introduced by randomness in the developed grain pattern. Second, the conversion of a film stored optical image to an electrical form, through illumination and photon detection, results in photoelectronic noise. This occurs, for example, in a microdensitometer scan across a film region. Photoelectronic noise is also present in digital neutron radiography systems employing television cameras. Third, electronic noise arises from electron random thermal motions during amplification of the electrical signal. Unlike X-ray radiography, neutron radiographic quantum noise, due to statistical variations in the ray image, is a fourth noise source that is usually small relative to these other three sources. The three important noise sources (B.4, B.11) can be briefly reviewed and characterized as follows:

Film-grain noise: Randomness in the spatial deposition of developed grains arises (1) from the randomness in size, shape and separation of silver halide grains on the film prior to exposure, and (2) from the randomness in silver deposition between similar grains under similar exposures. To a good approximation, film-grain optical density noise is characterized by a Gaussian probability density function with a standard deviation that increases with the mean. Film-grain noise is thus signal dependent.

Photoelectronic noise: Photoelectronic noise is also signal dependent, arising from the statistical variations in both the photons incident on light sensors and in the photoelectronic conversion process of these sensors. It is usually modelled as being Poisson or Gaussian distributed, with a standard deviation proportional to the square root of the mean.

Electronic noise: Electronic noise arises in resistive circuit elements and is usually modelled as a zero-mean Gaussian process, with equal power at all frequencies. It is thus commonly referred to as "Gaussian white noise".

An effective method for evaluating total system noise is the Wiener spectrum, being the power spectrum of the overall noise component, although such analysis is beyond the scope of this section. A brief characterization of the summation of the three Gaussian noise sources described above is simply that the individual variances sum, so that the overall noise standard deviation, σ_{tot}, is given by

$$\sigma_{tot} = \sqrt{\sigma_{fg}^2 + \sigma_{pe}^2 + \sigma_e^2} \tag{9.31}$$

where σ_{fg}, σ_{pe} and σ_e are the standard deviations for film-grain, photoelectronic and electronic noise, respectively.

9.8 NOISE SMOOTHING

Just as differentiation tends to magnify the noise accompanying a signal, integration tends to reduce noise. This becomes obvious under the recognition that integration is effectively an averaging procedure. If at each point a noisy signal is averaged with neighboring points, then the noise should be reduced in the resulting signal. The mathematical construct through which this averaging is performed is the convolution integral and the procedure is referred to as "convolution filtering", since noise is effectively filtered out of the signal. The input signal, I, may possess either spatial or temporal dependence, and the weighting employed in averaging is determined by the shape of the filter function, F. The smoothed output signal, 0, is thus given generally by

$$0 = I * F \tag{9.32}$$

In frequency terms, the filter function possesses a low response at high frequencies and is thus referred to as a "low-pass filter". A problem encountered with the non-negative filter functions frequently employed in noise smoothing is the simultaneous smoothing of the noise-free signal, to an extent dependent on the filter's modulation of the frequencies comprising the noise-free signal. This problem is illustrated in Fig. 9.11, depicting the filtering of a noisy edge response function with a flat filter function. In such cases, image sharpness is compromised by noise reduction.

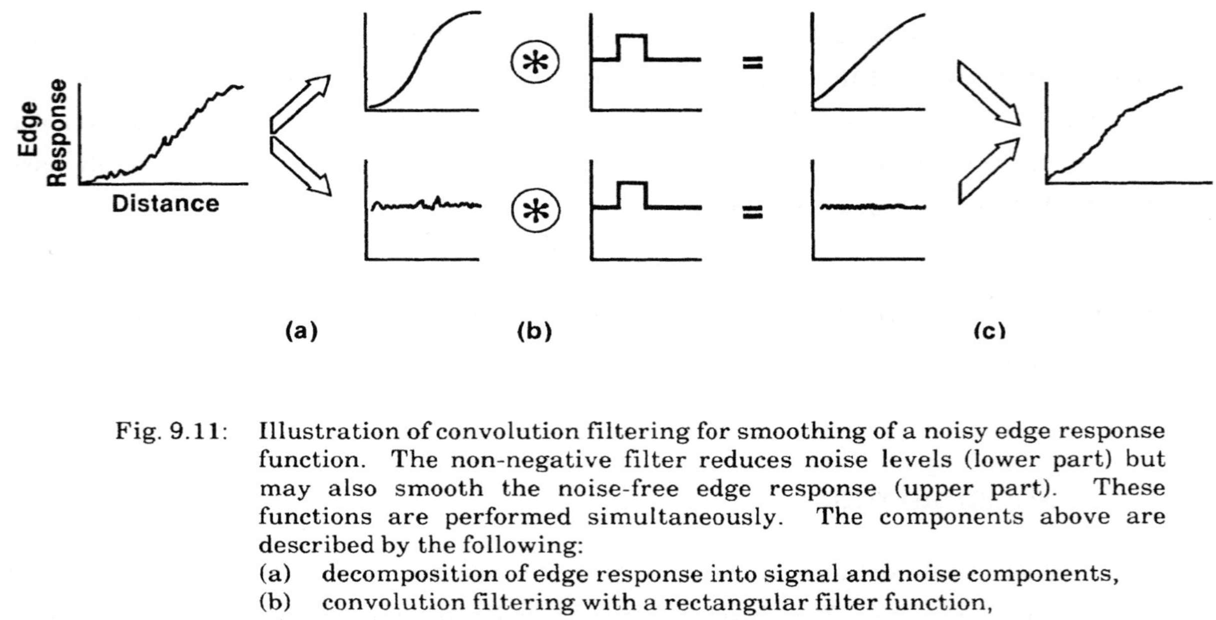

Fig. 9.11: Illustration of convolution filtering for smoothing of a noisy edge response function. The non-negative filter reduces noise levels (lower part) but may also smooth the noise-free edge response (upper part). These functions are performed simultaneously. The components above are described by the following:
(a) decomposition of edge response into signal and noise components,
(b) convolution filtering with a rectangular filter function,
(c) original edge response now smoothed.

There are several methods by which convolution filtering can be performed. The image can be first digitized and then discrete convolution can be performed. If the radiograph is being scanned with a microdensitometer, then the microdensitometer window serves as a flat filter function. Care must be exercised to ensure window dimensions that are adequate to smooth film-grain noise without unduly smoothing the desired signal. Even if film-grain noise is eliminated, electronic and photoelectronic noise may continue to exist. The microdensitometer scan itself may be digitized, thereby enabling digital filtering. Note that visual smoothing of a microdensitometer scan, while subjective, is actually a form of convolution filtering since a consideration of neighboring values presumably enters into the judgement as to the noise-free signal value at each point.

X. Neutron Radiographs

The following provide a sampling of neutron radiographs from various contributors and are intended to suggest the specialized features and range of application of neutron radiography. Further details about the radiographs are available from the contributors.

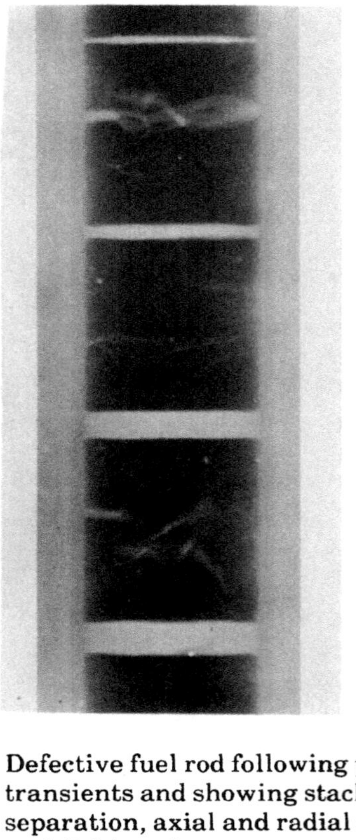

Fig: 10.1: Defective fuel rod following power transients and showing stack separation, axial and radial cracks, and dishing closure. (J. Markgraf, Joint Research Centre, Petten, The Netherlands).

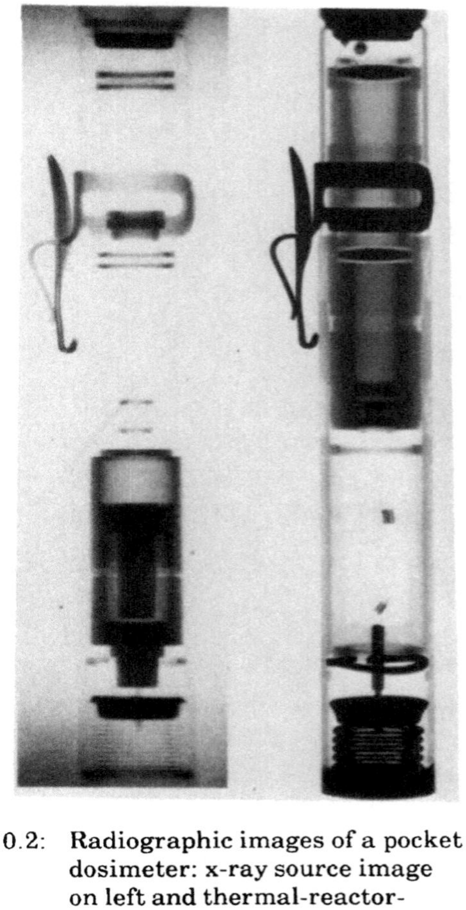

Fig. 10.2: Radiographic images of a pocket dosimeter: x-ray source image on left and thermal-reactor-source neutron image on right. (M. Hawkesworth, University of Birmingham, England).

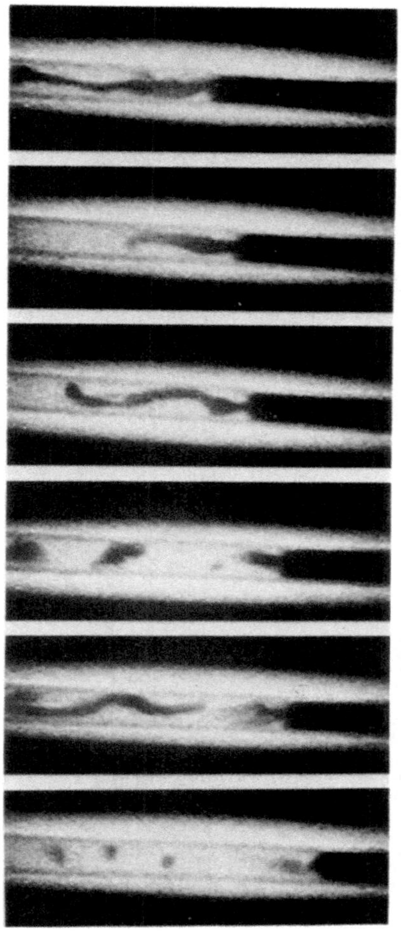

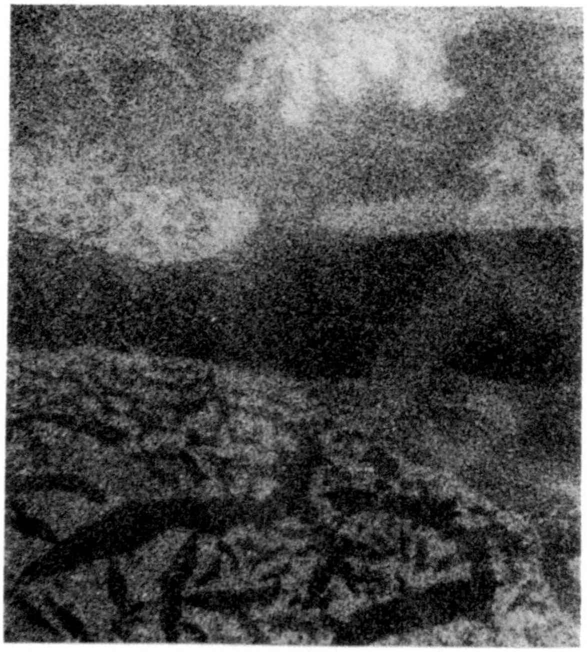

Fig. 10.4: Uneven distribution of adhesives between metals. (H. Rauch, Atominstitut, Austria).

Fig. 10.3: Real-time video images of two-phase flow; the sequence illustrates how excess liquid is entrained by steam generated at an advancing quench front. (P. Schofield, AERE Harwell, England).

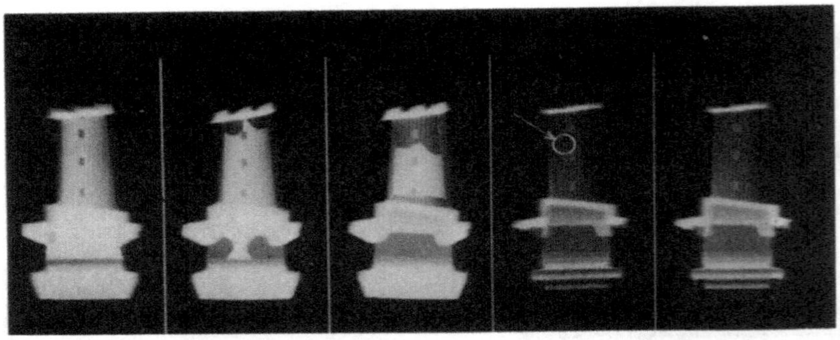

Fig. 10.5: Gas cooled aircraft turbine blades with various amounts of residual core material tagged with 1.5% Gd_2O_3. (R.L. Newacheck, Aerotest Operations, Inc., USA).

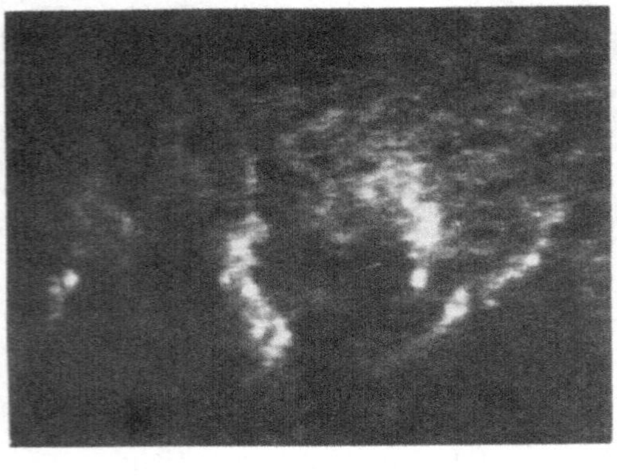

Fig. 10.6: Non-uniformity of B_4C distribution in boral composite, (N.C. Jain, Bhabha Atomic Research Centre, India).

Fig. 10.7: Miniature motor radiographed by track-etch methods using a $^{10}B_4C$ converter. (G. Matsumoto, Nagoya University, Japan)

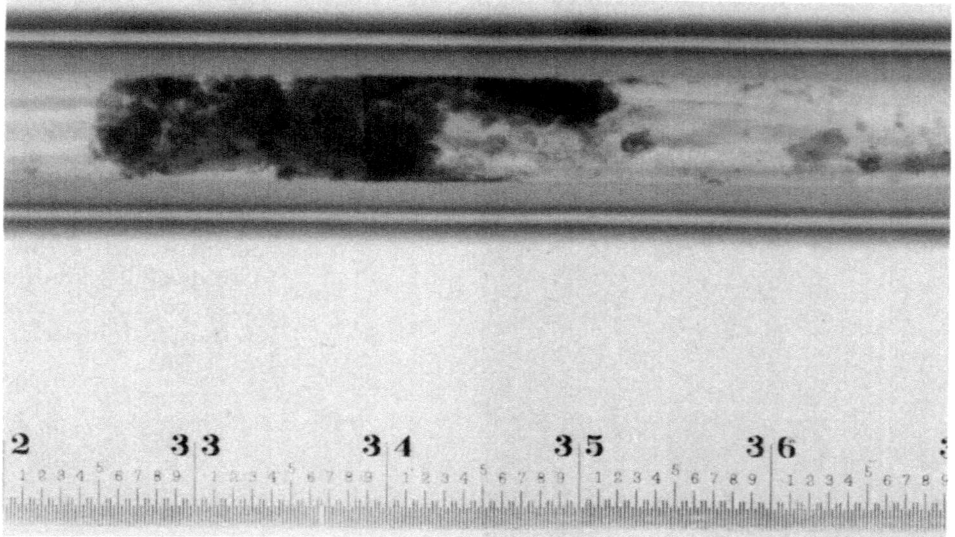

Fig. 10.8: Fuel melt resulting from transient tests. (W.J. Richards, Argonne National Laboratory West, USA).

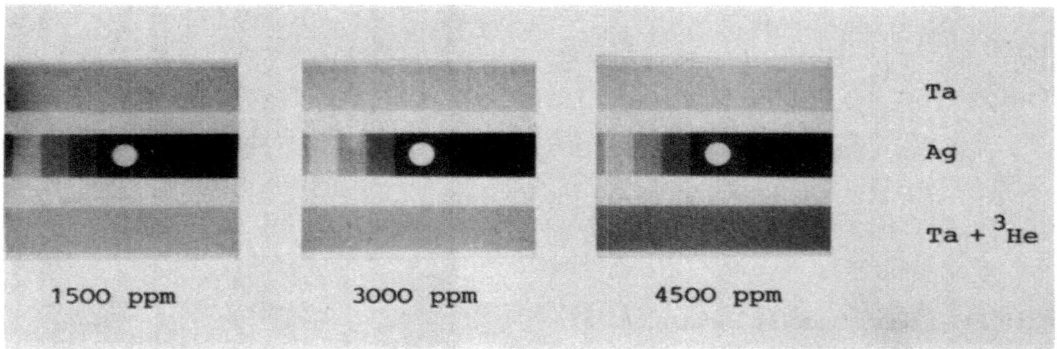

Fig. 10.9: Observation of ^{3}He production after the decay of tritium in a tantalum sample after 75, 350 and 530 days. The Ag-steps serve as reference densities. (H. Rauch, Atominstitut, Austria).

Fig. 10.10: Side view of aluminum fatigue specimen with arrow pointing to an alcohol-gadolinium oxide slurry to outline crack front. (H. Berger, Industrial Quality, Inc., USA).

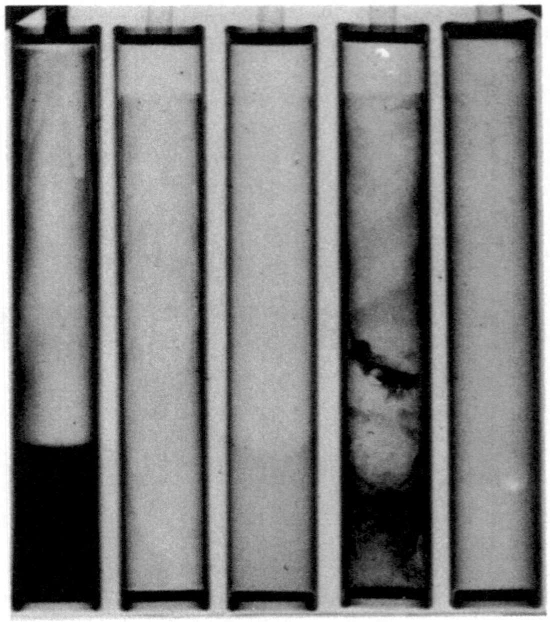

Fig. 10.11: Test models of heat-pipes containing various working fluids; left-to-right: H_2O, Thermes, Na, Na + Li (10%), Iso-octane (G. Matsumoto, Nagoya University, Japan).

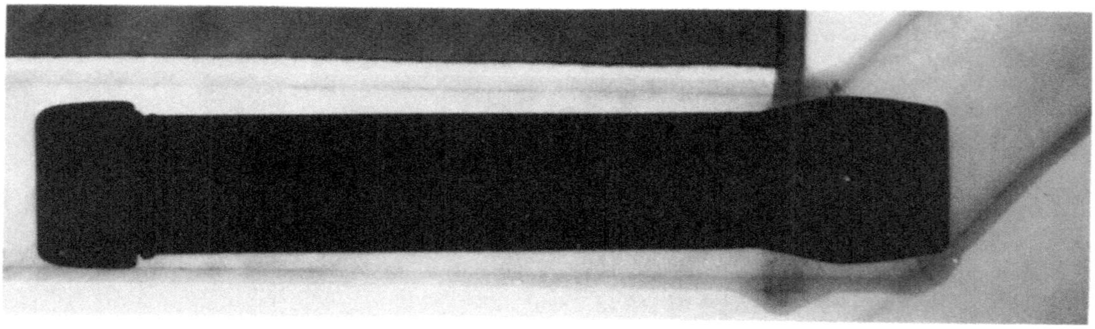

Fig.10.12: Pneumatic sample holder improperly stuck in a radioactive pipe at a reactor installation. (W.L. Whittemore, GA Technologies, Inc., USA).

Fig.10.13: Zircaloy-clad unirradiated fuel elements containing $(Th,Pu)O_2$ showing uniformity of plutonium distribution. (A.M. Ross, Chalk River Nuclear Laboratories, Canada).

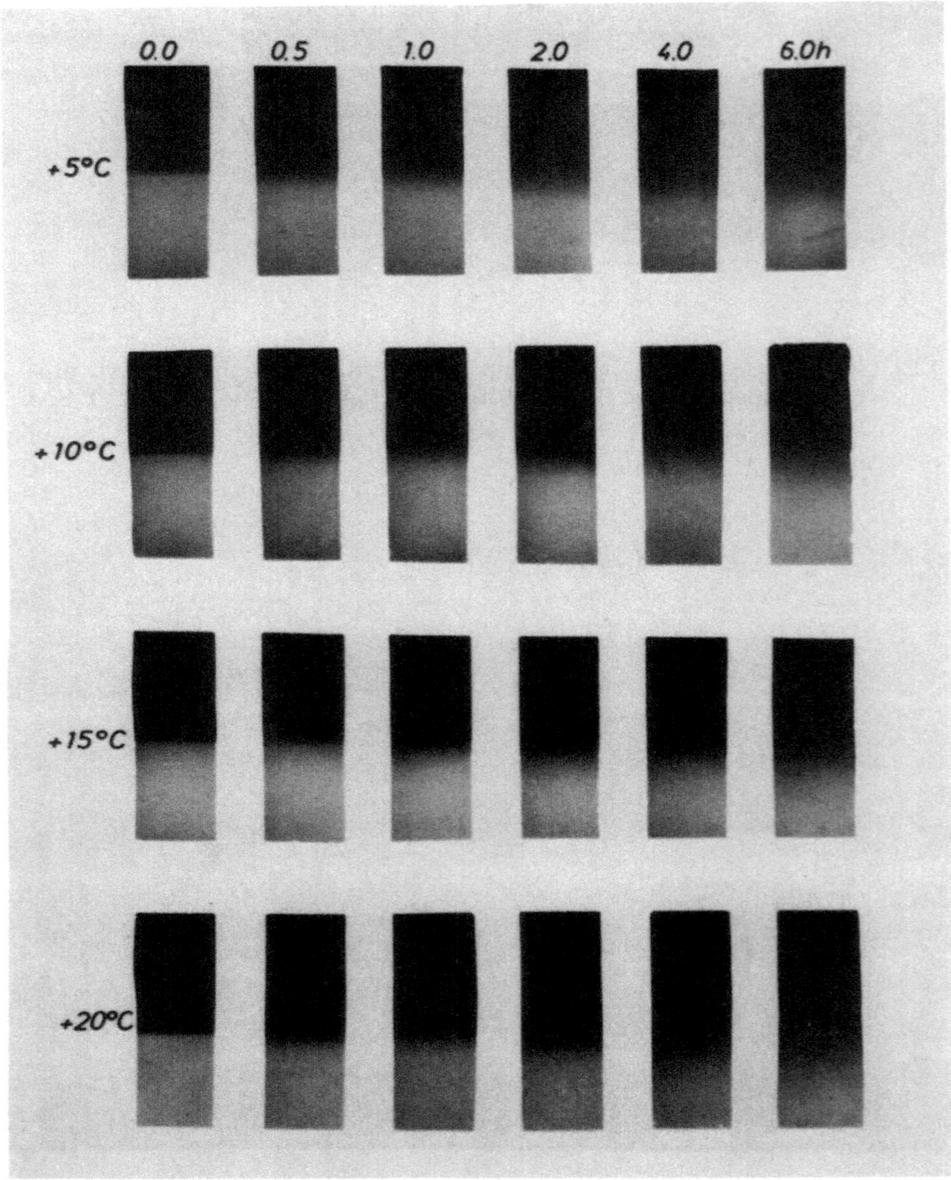

Fig.10.14: Diffusion between H_2O and D_2O as a function of time and temperature (H. Rauch, Atominstitut, Austria).

Fig.10.15: Epoxy foam filled electronic filters: only the second from the right meets prescribed quality assurance standards. (R.L. Newacheck, Aerotest Operations, Inc., USA).

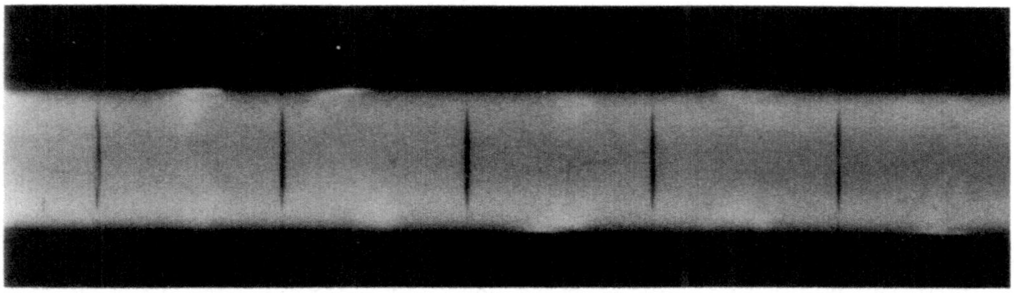

Fig.10.16: Fuel pin showing hydrides on the inside of cladding due to inleakage of water. (U. Bergenlid, Studsvik Energitechnik AB, Sweden).

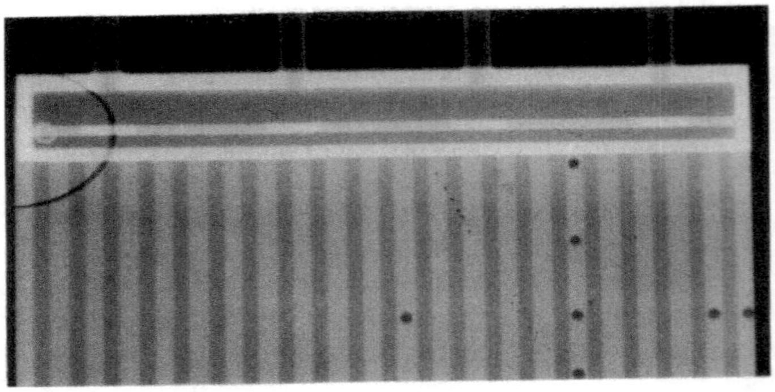

Fig.10.17: Brazed manifold containing an unwanted O-ring in the black circled area. (R.L. Newacheck, Aerotest Operations, Inc., USA).

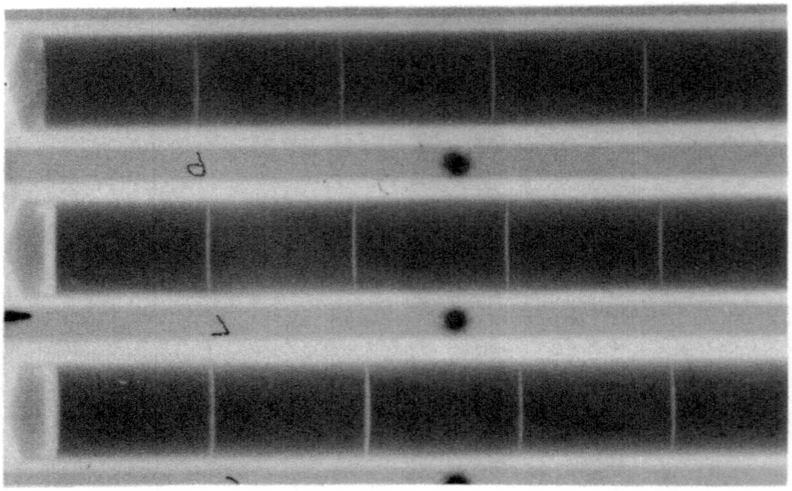

Fig.10.18: Natural UO_2 fuel after two years of reactor exposure. (A.M. Ross, Chalk River Nuclear Laboratories, Canada).

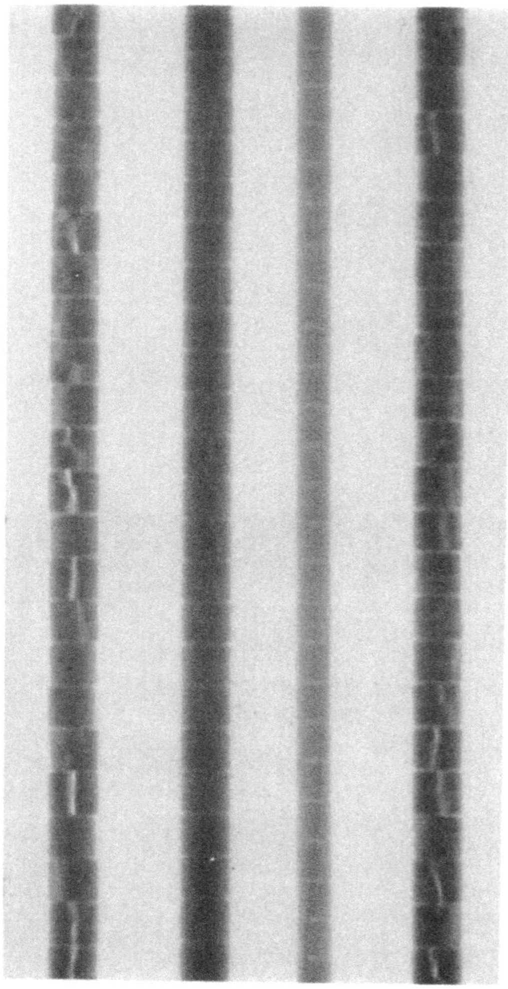

Fig.10.19: Epithermal neutron radiograph of carbide fuel pins showing cracking and loss of alignment. (W.J. Richards, Argonne National Laboratory West, USA).

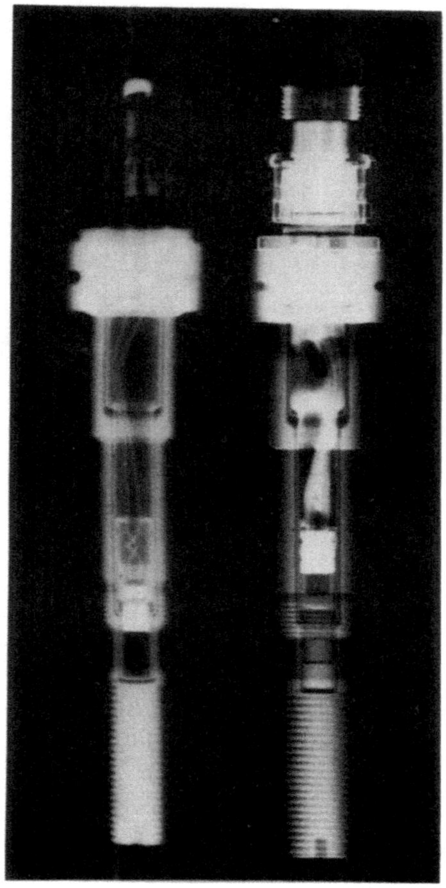

Fig. 10.20: Comparison between a x-ray (left) and a neutron radiograph (right) of an explosive bolt. (R.L. Newacheck, Aerotest Operations, Inc., USA).

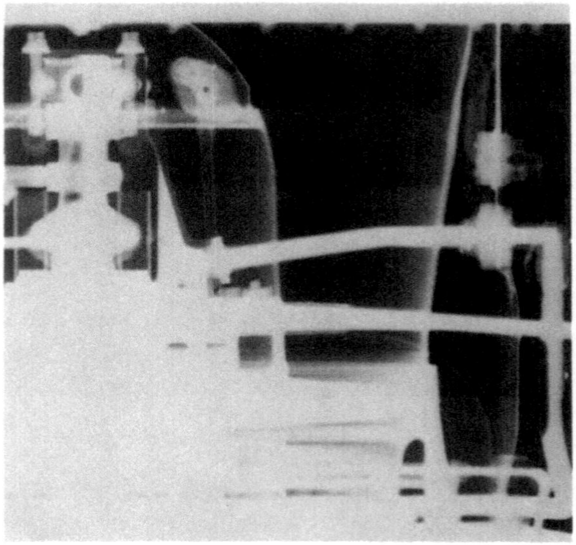

Fig. 10.21: Oil flow in an aircraft gas turbine engine while operating (upper) and shutdown (lower). (P. Schofield, AERE, Harwell, England).

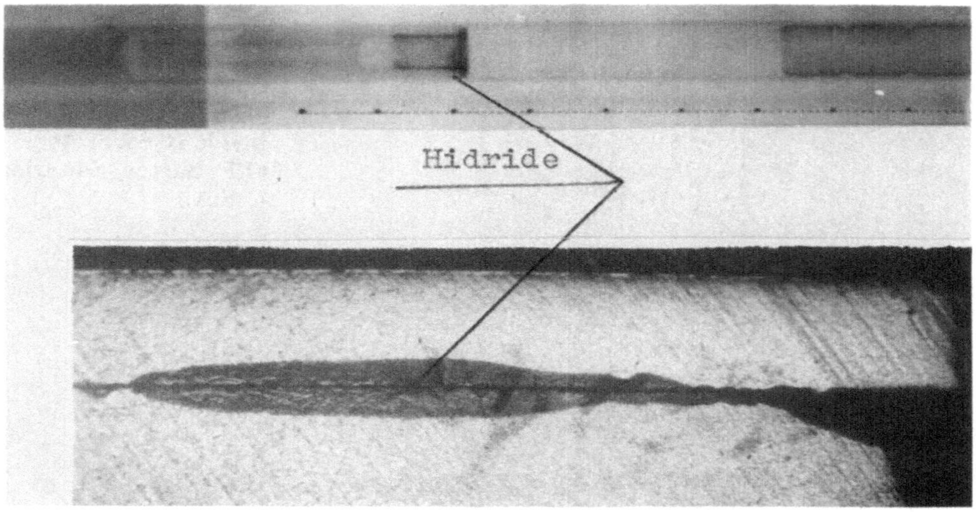

Fig. 10.22: Hidriding in a fuel element following introduction of water. (V.T. Korneev, Kurchatove Institute for Atomic Energy, U.S.S.R.).

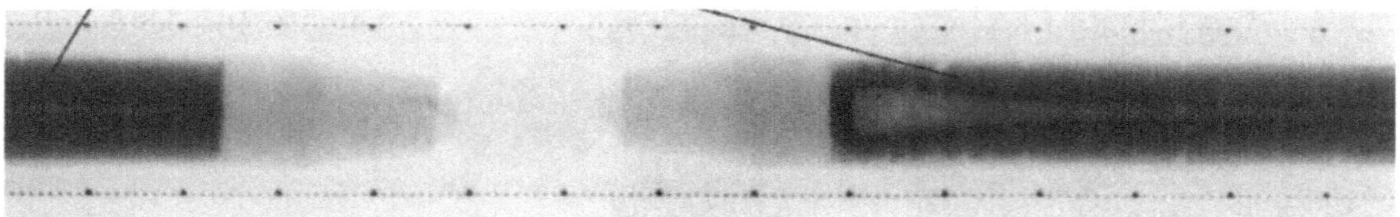

Fig. 10.23: Neutron radiograph of fuel element following 20,000 hours of reactor burnup. (V.T. Korneev, Kurchatove Institute for Atomic Energy, U.S.S.R.).

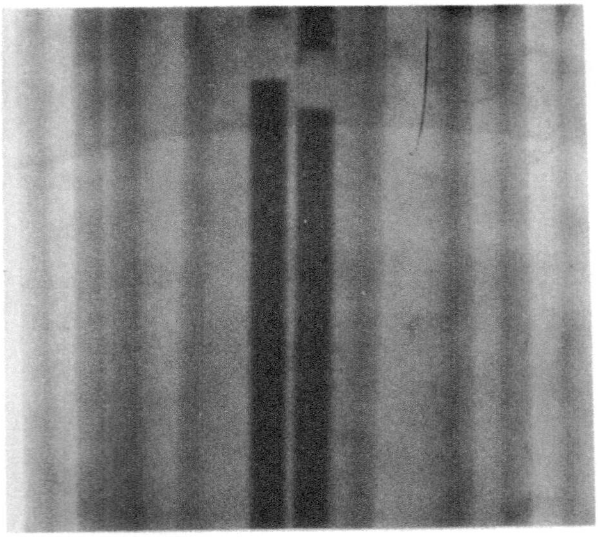

Fig. 10.24: Cold neutron radiograph of simulated nuclear fuel inside 10 cm of steel. (J.P. Barton, San Diego, U.S.A.)

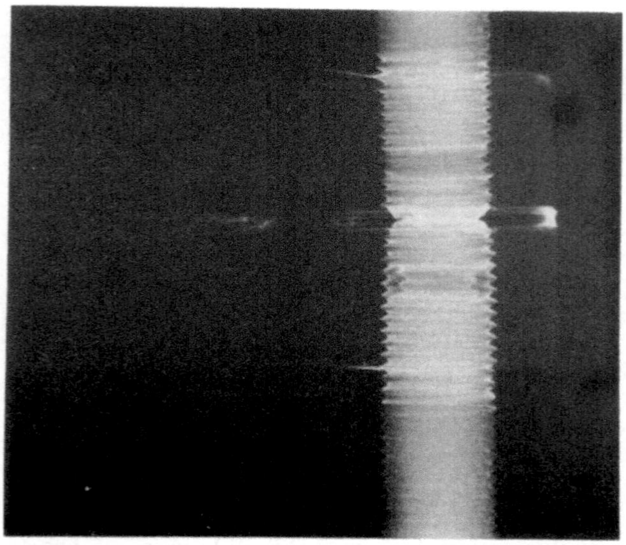

Fig. 10.25: Aluminum double cantilever beam showing hydrogenous deposits along crack. (H. Berger, Industrial Quality, Inc., USA)

Fig. 10.26: Thermal neutron radiographs of electronic filters being inspected for potting. (M.P. Butler, McMaster University, Canada).

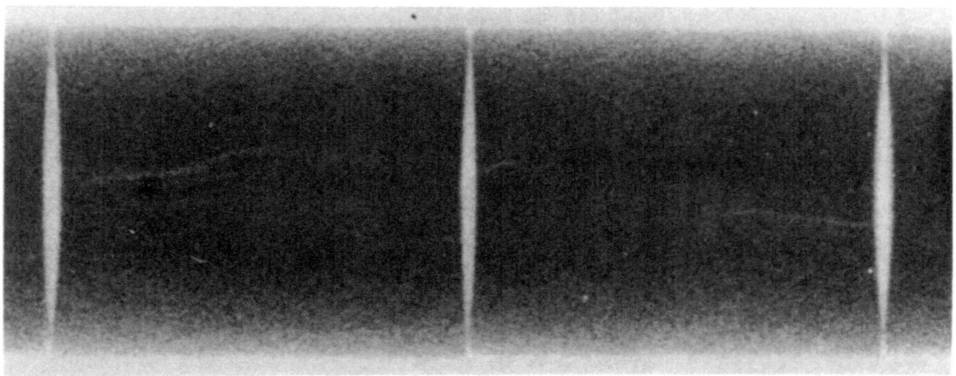

Fig. 10.27: Non-defective PWR fuel rod following power transient and showing detail of dishing volume (J. Markgraf, Joint Research Centre Petten, The Netherlands).

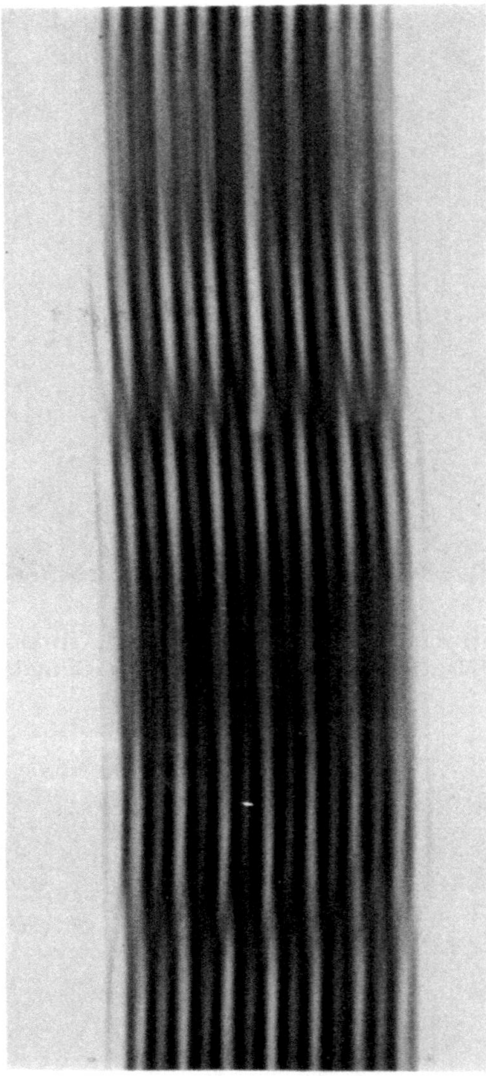

Fig. 10.28: Resonance neutron radiograph of a 91 pin fuel bundle. (J.P. Barton, San Diego, U.S.A.).

XI. Bibliography and References

The following lists publications generally available in the open literature relevant to our chosen emphasis on neutron radiography and which may be useful for further study. Those in the first section are listed chronologically in order to illustrate the conceptual and technological evolution of neutron radiography; those in the second section provide relevant information on special aspects of imaging and nuclear science. The identification numbers correspond to those used in the text for reference purposes.

SECTION A.

1940's

A.1 O.Z. Peter, "Neutronen-Durchleuchtung", Naturforschung, 1, 557 (1946).

A.2 H. Kallman, "Neutron Radiography", Research, 1, 254 (1948).

1950's

A.3 J. Thewlis, "Neutron Radiography", Brit. J. Appl. Phys., 7, 345 (1950).

1960's

A.4 H. Berger and W.N. Beck, "Neutron Radiographic Inspection of Radioactive Irradiated Reactor Fuel Specimens", Nucl. Sci. Eng., 15, 441 (1963).

A.5 J.P. Barton, "Some Possibilities of Neutron Radiography", Phys. Med. Biol., 9, 33 (1964).

A.6 J.P. Barton, "Contrast Sensitivities in Neutron Radiography", Applied Materials Research, 90 (1965).

A.7 H. Berger and W.N. Beck, "A Thermal-Neutron Television System for Postirradiation Annealing Studies", Trans. Am. Nucl. Soc., 8, 73 (1965).

A.8 H. Berger, Neutron Radiography: Methods, Capabilities and Applications, Elsevier Publ. Co., N.Y. (1965).

A.9 W.A. Carbiener, "Nondestructive Examination of Radioactive Material using Neutron Radiography", Nucl. Appl., 2, 468 (1966).

A.10 T. Inouye and K. Ogawa, "A Method of Determination of Thermal Neutron Beam Collimation", Nucl. Inst. Meth., 42, 309 (1966).

A.11 G. Oster, "A Young Physicist at Seventy: Hartmut Kallmann", Physics Today, 19, 51 (1966).

A.12 D.C. Cutforth and V.G. Aquino, "Neutron Radiography in the EBR-II Fuel Cycle Facility using an Isotopic Neutron Source", Trans. Am. Nucl. Soc., 10, 442 (1967).

A.13 J.R. Shoptaugh, Jr., and W.L. Whittemore, "High Resolution Neutron Radiography Using Pulsed and Steady-State Reactor Sources", Trans Am. Nucl. Soc. 10, 72 (1967).

A.14 B. Feigl and H. Rauch, "Der Gd-Neutronenzaehler", Nucl. Instr. Meth., 61, 349 (1968).

A.15 M.R. Hawkesworth and J. Walker, "Review: Radiography with Neutrons", J. Mat. Sci., 4, 817 (1969).

A.16 M.R. Hawkesworth, "Radiography with Low Intensity Neutron Beams: Some Reciprocity Characteristics of Film used with Light-emitting Intensifying Screens", J. Sci. Inst. (J. Phys. E.), 2, 673 (1969).

A.17 W.L. Whittemore, Fluxes, Beam Intensity, Collimation and Resolution for Neutron Radiography,, Gulf General Atomic Inc., San Diego, CA, USA, GA-9472, (1969).

1970's

A.18 S.J. Basham, D.R. Grieser and J.W. Ray, "Dimensional Measurements of Cylindrical Specimens using Neutron Radiography". Mater. Eval., 26, (6), 140 (1970).

A.19 A.A. Harms and B.K. Garside, "An Analytical Model of the Recording Process in Neutron Radiography", Trans, Am. Nucl. Soc., 13, 542 (1970).

A.20 H. Berger, "Neutron Radiography", Annual Rev. of Nucl. Sci., 21, 335 (1971).

A.21 B.K. Garside and A. A. Harms, "The Detection Process in Neutron Radiography", J. Appl. Physics, 42, 5161 (1971).

A.22 A.A. Harms and M.H. Younis, "Beam Divergence Contributions to Image Spread in Neutron Radiography", Trans. Am. Nucl. Soc. 14, 526 (1971).

A.23 I. Panaitescue, "Assessing the Quality of a Radiographic Image", Mater. Eval., 29, 153 (1971).

A.24 R.S. Matfield, "Neutron Radiography", Atom, 174, 1 (1971)

A.25 R.W. Parish, "Neutron Radiography", Engineering, 211, (33) (1971).

A.26 J.A. Rau and W.L. Parker, "Measurements of Antiscatter Grid Effectiveness in Thermal-Neutron Radiography of Hydrogenous Materials", Trans. Am. Nucl. Soc., 14, 533 (1971).

A.27 R.H. Bossi, J.L. Carson and C.N. Jackson, Jr., "The Modulation Transfer-Function and Effective Focal Spot as Related to Neutron Radiography", Mat. Eval., 103 (1972).

A.28 J.P. Barton and M.F. Klozar, "Experiments with ^{241}Am-^{242}Cm-Be for Neutron Radiography", Mat. Eval., 236 (1972).

A.29 A.A. Harms and G.R. Norman, "The Role of Internal Conversion Electrons in Gadolinium Exposure Neutron Imaging", J. Appl. Phys., 43, 3209 (1972).

A.30 A.A. Harms, B.K. Garside and P.S.W. Chan, "The Edge-Spread Function in Neutron Radiography", J. Appl. Phys., 43, 3863 (1972).

A.31 A.A. Harms and T.G. Blake, "Densitometer-Beam Effects in High Resolution Neutron Radiography", Trans Am. Nucl. Soc., 15, 710 (1972).

A.32 M. Muellner and H. Jex, "Converter Thickness for Optimum Intensity in Neutron Radiography", Nucl. Inst. Meth., 103, 229 (1972).

A.33 D.C. Cutforth, "Dimensioning Reactor Fuel Specimens from Thermal Neutron Radiographs", Nucl. Tech., 18, 67 (1973).

A.34 A.A. Harms, T.G. Blake and J.P. Marton, "Neutron Imaging with Thin Gadolinium Converters", Nucl. Inst..Meth.,109, 253-255 (1973).

A.35 A.A. Harms and B.K. Garside, "Radiographic Image Resolution Based on Edge-Spread Function Analysis", Trans. Am. Nucl., Soc., 18, 62 (1974).

A.36 A.A. Harms and G. McCormack, "Isotopic Conversion in Gadolinium-Exposure Imaging", Nucl. Inst. Meth., 118, 583 (1974).

A.37 A.A. Harms and M.S. Moniz, "A Framework for Radiation-Conversion Track-Etch Imaging Systems Analysis", Nucl. Inst. Meth., 121, 477 (1974).

A.38 A.A. Harms and M.S. Moniz, "Track-Etch Imaging Efficiencies of the ^{10}B(n,α)^{7} Li and ^{6}Li (n,α) T Reactions", Nucl. Inst. Meth., 122, 567 (1974).

A.39 L.A. Thaler, "Measurement of Capsule Heat Transfer Gaps using Neutron Radiography", Mater. Eval., 32, (3), 57 (1974).

A.40 K. Tomii, "Neutron Radiography Experiments at JRR-4", J. Nucl. Sci. Tech., 11, 153 (1974).

A.41 A. Vary and K. J. Bowles, "Application of an Electronic Image Analyzer to Dimensional Measurements from Neutron Radiographs", Mater. Eval., 32, (1), 7 (1974).

A.42 C.L. Aseltine and R.A. Strich, "Time Resolved Neutron Radiography using a Fast Pulse Reactor", Nucl. Tech., 26, 107 (1975).

A.43 M.R. Hawkesworth, Ed., Radiography with Neutrons, British Nuclear Energy Society, London, England (1975).

A.44 M. Copic, D. Horvat, R. Ilic, M. Najzer and J. Rant, "Ueber Die Innere Unschaerfe Einiger Radiographischer und Neutronenradiographischer Pruefuerfahren fuer Aktive Stpaltstoff-Elemente", Materialpruefung, 18, 171 (1970).

A.45 J.C. Domanus, "Accuracy of Dimension Measurements from Neutron Radiographs of Nuclear Fuel Pins", Proc. Eighth World Conf. Nondestructive Testing, Cannes, France (1976).

A.46 H. Berger, Ed., Practical Applications of Neutron Radiography and Gaging, American Society for Testing of Materials, Philadelphia, PA, USA (1976).

A.47 G.S. Okawara and A.A. Harms, "Neutron Radiography of Fast Transient Processes", Nucl. Tech., 31, 250 (1976).

A.48 A.A. Harms and A. Zeilinger, "A New Formulation of Total Unsharpness in Radiography", Phys. Med. Biol., 22, 70 (1977).

A.49 A.A. Harms, "Conversion Efficiency Enhancement in Neutron Radiography", Trans. Am. Nucl. Soc., 26, 156 (1977).

A.50 W.J. Hartmann and A.A. Harms, "Directional Effects in Track-Etch Imaging", Nucl. Inst. Meth., 140, 137 (1977).

A.51 D. Twersky, Atomic Energy Review, (Vol. 15, #2) International Atomic Energy Agency, Vienna, Austria (1977).

A.52 A.A. Harms, M. Heindler and D.M. Lowe, "The Physical Basis for Accurate Dimensional Measurements in Neutron Radiography", Mat. Eval., 36(5), 49 (1978).

A.53 A.A. Harms, M.P. Butler and J.C. Osuwa, "Non-Symmetric Unsharpness in Radiography", Trans. Am. Nucl. Soc. 32, 208, (1979).

A.54 K. Neelekantan and V. Subramanian, "A Simple Edge Enhancer for Neutron Radiography", Nucl. Instr. Meth., 158, 399 (1979).

1980's

A.55 P. von der Hardt and H. Roettger, Neutron Radiography Handbook, D. Reidel Publ. Co., Dordrecht, Holland (1981).

A.56 Y. Segal, A. Gutman, A. Fishman and A. Notea, "Point Spread Functions due to Neutron Scattering in Thermal Neutron Radiography of Aluminum, Iron, Zircaloy and Polyethylene Objects", Nucl. Instr. Meth., 197, 557 (1982).

A.57 D.R. Wyman and A.A. Harms, "Two-Dimensional Radiographic Imaging Analysis", J. Nondest. Eval., 3 (4), 221 (1982).

A.58 J.P. Barton and P. von der Hardt, Eds., Neutron Radiography, (Proc. First World Conf. Neutron Radiography), D. Reidel Publ. Co., Dordrecht, Holland (1983).

A.59 D.R. Wyman and A.A. Harms, "Neutron Image Blurring in Two Dimensions", Trans. Am. Nucl. Soc. 44, 206 (1983).

A.60 J.C. Domanus, Ed., Reference Neutron Radiographs of Nuclear Reactor Fuel, D. Reidel Publ. Co., Dordrecht, Holland (1984).

A.61 D.R. Wyman and A.A. Harms, "System Transfer Function Applications in Neutron Radiographic Object Scattering", Nucl. Sci. Eng. 88(4), 522 (1984).

A.62 D.R. Wyman and A.A. Harms, "The Radiographic Edge Scattering Distortion", J. Nondest. Eval., 4(2), 75 (1984).

SECTION B.

B.1 W.C. Roentgen, "On a New Kind of Rays", Nature, 53, 274 (1896).

B.2 J. Chadwick, "Possible Existence of a Neutron", Nature, 129, 312 (1932).

B.3 R.A. Brooks and G. DiChiro, "Principles of Computer-Assisted Tomography in Radiographic and Radioisotopic Imaging", Phys. Med. Biol., 21, 689 (1976).

B.4 H.C. Andrews and B.P. Hunt, Digital Image Restoration, Prentice-Hall, Englewood Cliffs, New Jersey (1977).

B.5 R.H. Herz, The Photographic Action of Ionizing Radiations, Wiley-Interscience, New York (1969).

B.6 G.B. Arfken, Mathematical Methods for Physicists, Academic Press, New York (1970).

B.7 E.W. Marchand, "From Line to Point Spread Function: The General Case", J. Opt. Soc. Am., 55, 352 (1965).

B.8 T.S. Curry III, J.E. Dowdy and R.C. Murry, Jr., Christensen's Introduction to the Physics of Diagnostic Radiology, 3rd Edition, Lea and Febiger, Philadelphia, Pennsylvania (1984).

B.9 R. Halmshaw, Physics of Industrial Radiology, American Elsevier, New York (1966).

B.10 J.J. Duderstadt and L.J. Hamilton, Nuclear Reactor Analysis, John Wiley and Sons, New York (1976).

B.11 K.R. Castleman, Digital Image Processing, Prentice-Hall, Englewood Cliffs, New Jersey (1979).

B.12 G.I. Bell and S. Glasstone, Nuclear Reactor Theory, Van Nostrand Reinhold, New York (1970).

B.13 H.E. Johns and J.R. Cunningham, The Physics of Radiology, 3rd Edition, Charles C. Thomas, Springfield, Illinois (1974).

B.14 L.L. Carter and E.D. Cashwell, Particle Transport Simulation with the Monte Carlo Method, USERDA Technical Information Center, Oak Ridge, Tennessee (1975).

B.15 R.N. Bracewell, The Fourier Transform and Its Applications, McGraw-Hill, New York (1965).

B.16 K. Rossmann, "Point Spread-Function, Line Spread-Function and Modulation Transfer Function", Radiology, 93, 257 (1969).

B.17 C.E. Metz and K. Doi, "Transfer Function Analysis of Radiographic Imaging Systems", Phys. Med. Biol., 24, 1079 (1979).

B.18 P. Spiegler and A. Norman, "The Total Unsharpness in Radiography", Phys. Med. Biol., 18, 884 (1973).

Subject Index

<u>GPSR Compliance</u>

The European Union's (EU) General Product Safety Regulation (GPSR) is a set of rules that requires consumer products to be safe and our obligations to ensure this.

If you have any concerns about our products, you can contact us on ProductSafety@springernature.com

In case Publisher is established outside the EU, the EU authorized representative is:

Springer Nature Customer Service Center GmbH
Europaplatz 3
69115 Heidelberg, Germany

Batch number: 10593311

Printed by Printforce, the Netherlands